WAR AND WARFARE
IN THE TWENTIETH CENTURY

To Paul, Sharon and William

War and Warfare
in the Twentieth Century

JIM STORR

Howgate Publishing Limited

First published in 2025 by
Howgate Publishing Limited
Station House
50 North Street
Havant
Hampshire
PO9 1QU
Email: info@howgatepublishing.com
Web: www.howgatepublishing.com

British Library Cataloguing-in-Publication Data
A catalogue record for this book is available from the British Library

ISBN 978-1-912440-57-3 (pbk)
ISBN 978-1-912440-56-6 (hbk)
ISBN 978-1-912440-70-2 (ebk - EPUB)

CONTENTS

FIGURES

INTRODUCTION

The past is our only good guide to the events of the future, but it is an imperfect mirror. Furthermore, we cannot truly recreate the conditions of war, short of war itself. That is just too dangerous. Therefore the best way for military practitioners to study their profession is to study the history of war and warfare. Their profession is war*fare*: the conduct of war.

Hence aim of this book is to consider what we can learn from war and warfare in the twentieth century. 'What we can learn' is problematic, as we shall see. Therefore the book simply seeks to make observations and deductions. It aims to be educative and pragmatic. That is, the book seeks to help the practitioner. That is one of several reasons why it does not 'engage with the literature' nor 'discuss the discourse'. Instead, 'War and Warfare in the Twentieth Century' considers the historical evidence, such as it is, and works from there. That can be difficult. Much of historical 'fact' is to some extent subjective.

Studying the twentieth century is useful. It was the bloodiest century in history. War was widespread, and not just during the ten years of the two World Wars. Furthermore, the twentieth is the most recent century which we can consider as a completed period in the past. It is therefore the most immediately relevant. Additionally, it seems that only one book has previously considered war and warfare in the twentieth century as a whole.[1] 'War and Warfare in the Twentieth Century' is a thoroughly revised, and very largely rewritten, development of that book.

This Introduction considers definitions and then two ideas – concepts – which are used in the analysis within this book.

The military world is riddled with definitions. Armed forces produce service glossaries. Staff officers spend months writing, amending and gaining endorsement for them. The results are, largely, internally consistent. Larger organisations, such as NATO, expend much time and

1 Jim Storr, *The Hall of Mirrors: War and Warfare in the Twentieth Century*. (Warwick: Helion and Company, 2016.)

Term	Meaning	Remarks
War	Collective armed violence for political purposes.	See also 'politics.'
Warfare	The conduct of war.	
Politics	The means by which power is brokered in a society.	Power relates to authority over legal, fiscal, economic and sometimes cultural issues. It includes control over the acceptable use of violence within the society.
Strategy	The application of military resources at the national level. *Grand strategy* is the political direction of a war or conflict; or preparations for such in peacetime. *Military strategy* is the translation of grand strategy into military activity.	The alternative usage, 'the art of planning military activity', is deliberately avoided.
The Operational Level	Describes, or refers to, military activities at the level of theatre or campaign. Hence 'operational art': the planning, sequencing and resourcing of campaigns and major operations.	The meaning of 'operational' as relating to military activity in general is largely avoided.
Tactics	The conduct of battles or engagements.	
Winning and Losing		Generally not used. 'Success' and 'failure' are normally used in their place. Hence for example 'strategic success'; 'operational failure'; and so on. See 'Victory'.

Table I-1 continued...

Victory	Victory is considered to be a declaratory political artefact, related to success in military activities.	
Command	The authority vested in an individual for the direction, co-ordination and control of military forces; or the exercise of that authority (that is, 'to command').	See also 'control'. The usage 'command and control' is obviously nugatory.
Control	Relates to the oversight, supervision and coordination of subordinates.	See also 'command'.

Figure I-1. Meanings Used within this Book

effort agreeing definitions between nations, services, or both. But it would be good if everything was that simple.

There are no 'definition police'. Many words relating to war and warfare have been in common use for centuries, and perhaps millennia. Common meanings have great virtue. Not least, they are generally commonly understood. However, common meanings can be very broad. They can have exceptions or be vague. This is a particular problem for the study of war and warfare. Alternative, or broad, meanings can give rise to ambiguity. So, some definitions are given here in figure I-1 for use within this book. They are generally narrow and specific, by design.

One consequence presents itself straight away. War is collective armed violence. Violence is damage intended to do harm.[2] Damage is inflicted by the use of weapons. Weapons are employed in battles and engagements. Critically, therefore, *all damage is tactical in the first instance.* Whether that damage has strategic, or operational, consequence depends entirely on the situation. It does not necessarily follow.

In sport, 'winning' and 'losing' are well defined. A soccer match is won by the side which scored the most goals by the end of the match.

2 *The Oxford English Dictionary* (OED).

Goals, and that 'end', are defined. In war, the meanings of winning and losing are both vague and subjective. Much the same applies to 'victory'. Hence the usage above.

Military operations (whether strategic, operational or tactical) can broadly be described as being either offensive or defensive. Within offensive operations there is a subset which typically is not identified as being separate in character. It is raiding. Raiding is a concept which deserves further attention.

Typically, aggressors attack and take possession of an area. They then attempt to retain that territory (or sea space, or whatever). Conversely, in raiding the aggressors attack, do damage and then move on. They may advance further or withdraw. But they make no effort to retain any area they seize. Raiding is an approach which can be employed at the strategic, operational or tactical levels.

Raiding can meet several purposes:

a. To gain economically (typically by removing and making use of resources).
b. To extract political benefit.
c. To deplete or destroy enemy supplies.
d. To live at the enemy's expense.
e. To compel battle.
f. To substitute for battle (for lack of alternative).[3]

We will consider raiding in various guises throughout this book. Here we will make one significant observation. All aircraft must land. They fly, do damage (or whatever), and return home. They *cannot* hold or retain any area. Land and naval warfare *may* include raiding as an offensive technique. Conversely *all* aerial warfare is raiding. That is worth stressing. *All aerial warfare is raiding*. Even air defence conducted by fighter aircraft consists of raids against attacking aircraft.

Turning to another concept, war is inherently complex. Tools can help us understand and manage complexity. It is useful to differentiate between three overall measures of performance when considering complex situations. They are economy, efficiency and effectiveness. See Figure I-2.

---

3 Archer Jones, *The Art of War in the Western World.* (London: Harrap, 1988), 679.

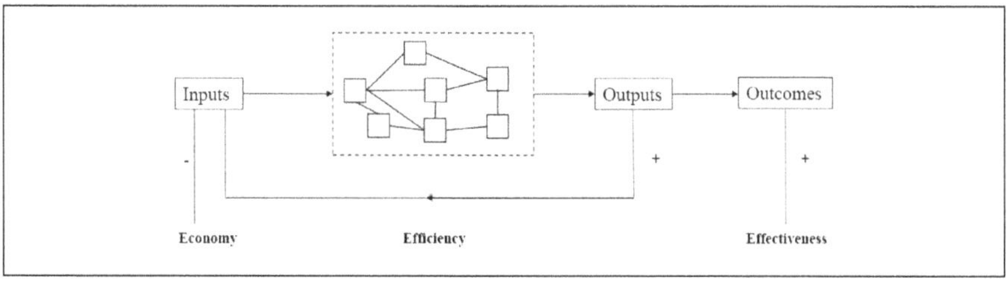

Figure I-2: Economy, Efficiency and Effectiveness[4]

A complex system or set of interactions – such as a battle or campaign – (shown here as a series of elements interacting within the dotted boundary) has inputs, outputs and outcomes:

- Outcomes are the effect of the outputs on the external environment
- Effectiveness relates to outcomes[5]
- Economy relates to minimising inputs
- Efficiency relates to the ratio of outputs to inputs.

In any complex system there may be no simple or direct relationship between input, output and outcome. Hence, in particular, the difference between output and outcome is highly important.

Take, for example, a jam and pickles factory. The inputs include sugar, water, fruit and vegetables. The main out*puts* are jars of jam or pickles. The main out*comes* are profit or loss for the owners. But say, for example, that the directors choose to use less fruit or sugar in the jam. That would mean less input. But the company may sell more of its jam, because it is now cheaper. Profits might go up. So, to repeat: there may be no simple or direct relationship between input, output and outcome. In a land battle, for example, outputs might be terrain taken or casualties inflicted or suffered. However the outcome is normally an issue of mission success or failure.

Perhaps more important, however, is to recognise that losses of personnel – casualties – are just that: losses. They are out*puts*, and negative (that is, losses rather than gains). They are not out*comes*. Clearly and understandably, casualties to ones' own forces should be avoided wherever

4 After von Bertalanffy. Jim Storr, *The Human Face of War.* (London, Continuum, 2009), 61-2.
5 OED.

possible. As we shall see, the difference between casualties as outputs, and success and failure as outcomes, is highly significant.

Having looked at definitions, raiding and complexity, we can now move on to the main purpose of this book. That is, to consider what we can observe or deduce from war and warfare In the twentieth century.

1

1900-1914

This chapter sets the scene in several ways. It describes the geopolitics, warfare, and wars of the period up to 1914. It then considers the origins of the Great War. By setting the scene it will allow us to see, in due course, how war and warfare changed through the century.

The world was a very different place. It was very euro-centric. There were no international political and diplomatic organisations, such as the United Nations (UN). In their absence international relations were dominated by individual statesmanship, diplomacy, and bilateral treaties (some of which were secret). In practice international relations largely centred on balance-of-power calculations between major European states.

Europe was home to five empires (the Russian, Ottoman, Austro-Hungarian, German and British) and one pseudo-empire (France). Almost all of Africa, and most of southern Asia, consisted of European colonies. Most of Latin America had only decolonised from Spain or Portugal in the previous century. Canada had recently moved, and Australia and New Zealand were about to move, from colonial to dominion status. Most of the Near East was divided into Turkish provinces. Almost all European countries were monarchies. France, again, was the main exception.

The USA had recently become the largest economy in the world. China had the next-largest Gross Domestic Product (GDP), but its vast population (and other factors) meant that it could not deploy its wealth outside of its own borders. However, in comparison, the combined economies of Russia, Germany, Austria-Hungary, France and Britain totalled more than twice that of the USA. That excludes the economies of Britain's empire (which brought Britain's GDP close to America's).

Two newcomers challenged the international order: Italy and Japan. Italy had been unified only a few decades before. Its foreign policy was aggressive and expansionist. Italy had interests in the Adriatic (and hence the Balkans) and Africa (hence Libya and Abyssinia (Ethiopia)). Japan was modernising, industrialising, belligerent and expansionist; especially in relation to Korea and Manchuria.

The structure of imperial governments was a particular problem. That applied to Japan as much as the European empires. In each case the emperor was in practice the only point at which all aspects of government met. Armies and navies were independent from each other (there were no 'defence' departments or ministries). Generals and admirals did not necessarily believe that strategy was a matter for civilian governments; including their foreign ministries. Military funding often dominated government expenditure. The funding of the Royal Navy, for example, was the single largest item of British government spending.

Yet emperors, and the people they appointed, were sometimes not very capable nor attractive people. The German Kaiser, Wilhelm II, was a difficult character who had alienated most of the other crowned heads of Europe; despite being related to most of them. Conrad von Hötzendorf, the Chief of the Austro-Hungarian General Staff, proposed declaring war on Serbia 25 times in 1913. War on Serbia may, or may not, have been appropriate; but von Hötzendorf's behaviour was not exactly measured or balanced.

The Industrial Revolution was moving on from its origins in coal and steel. New, chemical industries were emerging. The development of the Haber process, to produce ammonia, in Germany in 1909 was strategically significant. Without it, the Royal Navy could cut Germany off from the supply of guano it needed to produce explosives. By 1914 Germany was a world leader in chemical production. (Incidentally, that would give it an advantage in the production of poison gas.)

Naval theory was reasonably well developed. The works of the British Sir Julian Corbett and the American Captain Alfred Mahan were studied around the world. That led to some uniformity of tactical and operational thinking. Tactics centred on the Nelsonian line of battle dominated by the battleship, together with the relevant fleet manoeuvres and signalling.

Operationally, there was clear and mature understanding of issues such as control of the sea; the importance of maintaining a fleet in being (not least, to threaten the enemy's control of the sea); the use of the sea to move armies to other countries and continents; and war on trade (for which a reasonably well-developed body of international law existed).

The key technologies for warships, and particularly battleships, were the coal-fired steam turbine; the rifled, breach-loading gun; steel armour; and wireless signalling. The turbine had only been invented in 1884. Those four components were not combined into an integrated design until the British HMS *Dreadnought* was completed in 1906. Very few nations owned a fleet of battleships. No navy could afford to scrap its existing ships and replace them with Dreadnoughts directly. All major navies still had pre-dreadnought battleships in commission in 1914. Fire control was problematic, but the biggest guns were fairly accurate out beyond ten miles (16 km) or so.

Cruisers were next in importance. With some of the features of battleships, cruisers emphasized endurance and seakeeping. Navies used them primarily to patrol sea lines of communication. Since Britain had the largest and most far-flung empire, it had the most cruisers. It had 121 in 1914. The next eight nations had 178 between them. In fleet operations cruisers were used for scouting and screening.

Destroyers were optimised for high speed. They could reach up to 33 knots, compared with about 20 for a battleship. They had a light gun armament, but carried torpedoes which could sink the heaviest warship. They could not stay at sea in the worst weather conditions, and had limited range. They were generally employed in flotillas to support the line of battle.

Submarines were in their infancy. In 1914 they had very limited range and endurance. Admirals were broadly aware of their potential but even more aware of their current limitations.

On land, operational plans and tactics generally focussed on the offensive. In several armies there was an almost mystical belief in the power of the attack. Tactical reality, however, was dominated by the rifled bullet. It was cheap, accurate and lethal. It enabled infantrymen to deliver devastating fire out to ranges of several hundred yards. Rifled muskets firing conoidal bullets had been changing land tactics since the American Civil War, if not before. Armies were adjusting towards greater battlefield dispersion, the use of artillery to suppress defending infantry (enabling attacking infantry to assault), and a reappraisal of the effectiveness of mounted cavalry charges. The machinegun was beginning to appear on the battlefield in numbers. That would make the problem of attacking rifle-armed infantry even harder to solve.

Field artillery was evolving. Rifled barrels, fixed charges in brass cartridge cases, explosive shells and effective recoil mechanisms were key features. Fast-operating breaches and better recoil systems enabled much higher rates of fire. That, coupled to shrapnel shell, offered a way to suppress defending infantry. Heavier artillery was not particularly common. It was typically seen as a feature of siege warfare.

Commanders had been debating the role of cavalry for decades. They had not come to any firm conclusion. In the final months of the Franco-Prussian war 12 out of 16 Prussian cavalry charges had succeeded. The issue was not *whether* cavalry needed to adapt and evolve, but *how*. Cavalry was essential for scouting. Its use in screening depended on its ability to deliver dismounted fire. However, cavalry carbines were not particularly effective. The use of swords and lances was widely discussed. There was, perhaps, consensus that without their use cavalry was not capable of offensive action. Some doubted, however, whether offensive action could ever succeed against infantry.

Permanent fortifications were widespread. Many were sited to defend cities along European borders. Most consisted of rings of forts which would be linked together, in war, by trench systems. The most modern forts were built of reinforced concrete and could be very tough indeed. They were extremely expensive to build. In several cases fortresses were only a few decades old but, as we shall see, obsolete or obsolescent.

European armies were largely conscript. There were few regular units. The French Foreign Legion was a rare example. Many European countries could mobilise armies of hundreds of thousands of men in a few days. It would then typically take several more days to move them, by train, to their deployment areas. Training standards varied considerably between armies.

The American and British armies were different. They consisted of the equivalent of a few divisions of regular troops. They were often spread out across far-flung garrisons and had little or no formation training. In both cases the regulars could be reinforced by a few more divisions of volunteer reserves. Those reserves were often highly motivated but had relatively little training, and typically none at formation level.

Deployment to theatre was generally by steamship. Deployment within a theatre was by rail. From the railhead it was on foot or horseback. Canning and other forms of food preservation meant that rations were generally adequate. Inoculation and other preventative measures typically led to relatively low rates of sickness. Survival rates for the wounded were fairly good, due to developments in surgery. There were, of course, variations between armies. The Russian, Austro-Hungarian and Italian armies were not as well supported as their western counterparts.

In 1900 two major wars were underway. America had annexed the Philippines after the Spanish-American War of 1898, rather than recognize calls for independence. An insurrection broke out. America sent ships and land forces to the Philippines and initiated what soon became a counterinsurgency campaign. American troop strength reached about 126,000. They faced 60-80,000 insurgents. Fighting was sporadic and heavily one-sided. The Americans lost about 4,200 killed; the insurgents 16-20,000. However cholera broke out and perhaps 200,000 Filipino civilians died as a result. The insurgents' leader was captured in March 1901. The insurgents fought on for another year. There were atrocities on both sides. However, most American atrocities took the form of reprisals which were not, at the time, contrary to law.

The war was notable for the presence of veterans of the American Civil War. For example, General Arthur Macarthur took over as commander in chief in 1900. He had won the Medal of Honor at Missionary Ridge in 1863. General Douglas Macarthur, of Second World War fame, was his son.

The Second Anglo-Boer War had broken out in 1899. The discovery of gold had led to an influx of British civilians into the Transvaal and the Orange Free State. That led to violence between the immigrants and the local Afrikaner (originally Dutch) population. Afrikaner, or Boer, forces expelled British militia forces fairly quickly and then attacked British-held Natal and Cape Colony. The Boers initially mustered about 33,000 men; the British 13,000. The British suffered an early string of defeats. Nevertheless, they held the towns of Mafeking, Kimberley and Ladysmith, despite long sieges.

Much of the theatre consisted of vast, rolling plains. Cavalry was at a premium. The British initially had very little, whereas practically every Boer soldier was mounted on a horse or pony. The Boers fought on foot and were generally very good shots. Their rifle fire was highly effective out to long ranges. It took time and many casualties before the British learnt to disperse, use cover, and to employ wide outflanking movements. The British rushed both regular and reserve ('yeomanry') cavalry to South Africa. Several infantry battalions were converted into mounted rifles. The tide of the war turned. British force strength reached 480,000 whilst the

Boers never exceeded 80,000 men in the field. Britain's troops included Australian, New Zealand and Canadian contingents. Its cavalry performed well in the battles of Klip Drift and Elandslagte.

British forces captured the Boer capitals of Bloemfontein and Pretoria in the spring of 1900. They then waged a scorched earth campaign, moving Boer families into concentration camps, burning farms and rounding up Boer fighters. The much-reduced Boer forces signed a treaty in May 1902. The British Army's initial performance was heavily criticised. That led to a major series of reforms, which paid dividends in the Great War. As an example, its cavalry was re-equipped with rifles instead of carbines. Small arms training was improved significantly.

In 1904 Russia, expanding east to the Pacific, clashed with Japan (which was expanding into Manchuria). The resulting 18-month war was a major defeat for Russia. Waging war over 6,000km from Moscow was never going to be easy. On the outbreak of war Russia ordered its Baltic Fleet east. After a seven-month voyage around the Cape of Good Hope, the fleet was heavily defeated at the battle of Tsushima. Fought at ranges of about 5-10,000 yards, Tsushima resulted in three Russian battleships sunk for no loss to the Japanese. That night an attack by Japanese light forces resulted in the loss of four more Russian warships. The Russians surrendered the following morning. Of the 42 warships and auxiliaries that left Europe, only three were not sunk or surrendered.

Operations on land were, once again, dominated by rifle fire in defence. The campaign initially centred on the siege of the Russian Far East Fleet's harbour of Port Arthur and attempts to relieve it. Japanese attacks on entrenched Russian infantry were mostly bloody failures. Japanese siege artillery then enabled the capture of a key hilltop bastion from which they could direct fire on the Russian fleet. The Russian warships were then all sunk or scuttled in harbour. Port Arthur surrendered, and fighting shifted northwards. After a major battle at Mukden, the Russians were forced to withdraw. The defeat at Tsushima and a revolution throughout European Russia persuaded the Tsar to sue for peace.

With extensive trench lines, massed rifle fire, heavy artillery and some machineguns, the Russo-Japanese War displayed most of the features that would be seen on the Western Front in the Great War. Cavalry performance was disappointing on both sides. In the case of the Russians, and particularly Cossacks, that surprised western observers. The war was reported extensively in the west, but western armies generally chose to disregard its lessons.

The last major wars of the period took place in the Balkans. Nationalism, and particularly Serb nationalism, was a major factor. So was the decline of Ottoman Turkey. Relatively small nations mobilised large armies, which consisted mostly of rifle-armed infantrymen. Put simply, in the First Balkan war of 1912-13 Greece, Montenegro, Serbia and Bulgaria attacked and overran most of Turkey's European provinces. The Bulgarian army besieged Edirne for four months. Bulgaria had no heavy artillery and so could not defeat the fairly modern, German-designed fortifications. A Serbian siege train eventually arrived and tipped the balance. Edirne surrendered after two well-conducted Bulgarian night attacks.

The Second Balkan War broke out two months later. It lasted just five weeks. Again, in simple terms, Greece and Serbia considered that Bulgaria had profited too much (in terms of territory) from the First (Balkan) War. Romania joined Greece, Montenegro and Serbia. It wished to gain southern Dobruja from Bulgaria. Bulgaria was surrounded and heavily outnumbered. It lost heavily. It would therefore have understandable reasons to go to war against Serbia in 1915 and Romania in 1916.

Great Power involvement was a major factor in the Balkan Wars. Not least, the treaty ending the First Balkan War was signed in London. International interest in the Balkans was nothing new. It played a major part in the outbreak of the Great War 12 months later. Furthermore the Balkan states (and especially Serbia) now had experienced, veteran armies.

The assassination of the Austro-Hungarian Archduke Franz Ferdinand in late June 1914 was planned and directed by Serbian intelligence officers. It was aimed directly at the Austro-Hungarian monarchy: Franz Ferdinand was Emperor Franz Josef's nephew and heir. The assassination was either deliberately intended to start a war, or entirely reckless of the consequences.

Austria-Hungary's response was actually measured and deliberate. It sent an ultimatum to Serbia more than three weeks after the assassination. It was probably intended to be unacceptable to Serbia, and hence result in war. Serbia artfully rejected the ultimatum. Austria-Hungary declared war five days later.

'Balance of power' politics were already at play. Austria-Hungary had received undertakings of support from Germany (described below). A secret treaty of 1892 required Russia and France to mobilise if any of the Triple Alliance (of Germany, Italy and Austria Hungary) mobilised. Germany and Austria-Hungary did. The rest is recounted in Chapter 2.

Accounts of the origins of the Great War tend to stress structural issues such as Great Power politics and bilateral or multilateral treaties. They tend to gloss over human agency. We shall never really understand the motivation of the head of Serbian intelligence, Colonel Dragutin Dimitrijević (known as 'Apis'), because he was executed for treason in 1917. Similarly we will probably never know to what extent the Serbian prime minister, Nikola Pašić, was involved in the assassination plot. On the Austro-Hungarian side, Franz Ferdinand had intended to remove von Hötzendorf after the Autumn manoeuvres. Franz Ferdinand, however, was now dead.

The German government encouraged Austria-Hungary to proceed against Serbia with what has been described as a 'blank cheque'. Germany clearly had its own reasons, which would have seemed compelling at the time. Whatever they were, it is not reasonable to believe that the German government thought that it was signing up to four years of intense warfare to support the Habsburg monarchy. It was, however, inconceivable that Germany would offer to support Austria-Hungary without Kaiser Wilhelm's permission.

Many other people were involved in the outbreak of war. Some of them were just as obnoxious, if not more so. However, we can see from just these cases that individuals matter. Different people may well have acted differently. Structural issues alone do not account for the outbreak of the Great War.

France did not necessarily want war, but if war broke out France wanted it sooner rather than later. Its war plan was effectively a deployment plan which would give the commander in chief the greatest possible flexibility. The President and the Chief of the General Staff concealed the plan from the war minister and it was never discussed in parliament. However, the key problem was not a lack of parliamentary oversight. In practice France followed its offensive doctrine in 1914 without real consideration of the enemy's goals or plans.

Britain was well aware that Germany might overrun France in three weeks. After all, in 1871 it did. Germany, however, possessed the only fleet capable of challenging the Royal Navy. Britain's wealth depended on its maritime trade; London was the busiest port in the world; and the Thames lies opposite the Belgian ports. Britain's main interest was peace, because peace was good for trade. Peace and trade were not, however, likely to be maintained if a hostile fleet had use of the Belgian Channel ports. Thus Germany's plan to violate Belgian neutrality was naïve. It was strategy without political context.

What observations or deductions can we make about the pre-war period?

Firstly, nations were not particularly good at strategy. In the British case the government had failed, for several years, to either create or impose a united strategy on the General Staff and the Admiralty. In 1914 it did have a First Lord of the Admiralty (that is, the navy minister: Winston Churchill). However the post of Secretary of State for War (army minister) had been vacant for four months. Britain had middle-ranking and senior officers, and ministers, who were entirely capable of developing strategy. However it lacked an effective corporate method by which to do that. We will look at the consequences of poor strategic planning in Chapter 2.

Secondly, there is always unfinished business. At every moment several issues demand governments' attention. Some of those issues continue, to some extent in the background, through a major war and on afterwards. They can distract attention at critical moments. In Britain's case Ireland was such an issue. The (third) 'Home Rule' bill for Irish independence was being actively debated in Parliament right up until August 1914. It passed into law, but would not be enacted until after the War. Ireland was the reason why Britain did not have a Secretary of State for War. The incumbent, Colonel John Seely, had resigned. The 'Irish Question' was not resolved until 1922.

At sea, Dreadnought battleships were in service in some numbers by 1914. Admirals generally understood maritime blockade and convoying. Submarines had not yet made an impact. Hence the issue of commerce raiding by submarine had not arisen either. So convoy protection was not generally considered.

On land the devastating impact of the rifled, conoidal bullet was broadly recognised. However its consequences were not fully realised. Infantry tactics in the attack were in transition. It was thought that field artillery could have an important role in the attack. It would suppress the defence and hence support the infantry. The role of artillery in the defence was probably not fully understood. There was wide debate, and wide differences of opinion, as to the role and effectiveness of cavalry. In some quarters there was excessively rationalistic thinking. Cavalry attacks had to succeed, otherwise cavalry had no role. But cavalry had to have a role, so attacks would succeed. Or so it was thought, by some.

In 1914 many nations placed great faith in the power of permanent fortifications. As we shall see, in some ways they were right to do so. But not quite right enough.

Lighter-than-air craft were used by most armies and some navies. Armies used tethered balloons for observation: one had been present at Ladysmith in 1900. Navies were developing the use of dirigibles – airships – for fleet reconnaissance. To say that the use of aircraft was in its infancy, however, almost overstates the case.

Conversely subversion and espionage were moderately well developed. Espionage fiction was surprisingly common and popular. In Britain it was typified by Erskine Childer's immensely popular 'The Riddle of the Sands', published in 1903. (John Buchan's 'The 39 Steps', and others, were published during the Great War.) And, of course, what was the assassination of Franz Ferdinand if not subversive?

2

The Great War: Strategy

The First World War, or Great War, started when Austria-Hungary declared war on Serbia on 28 July 1914. By 23 August nine countries, including Japan, were at war. Chapter 2 looks at the Great War through strategy: that is, the conduct of war at the national level. It firstly considers the ends sought and the ways employed by the protagonists. The chapter then looks at the means deployed, the losses incurred and the outcome. Lastly it considers some of the more significant issues arising, and makes observations and deductions.

There are several misconceptions about the Great War. Much of that results from the historiography and how that has developed down to the present day. We need to unpick those misconceptions. That takes clear thinking and objectivity. Accordingly, some of this chapter may seem surprising.

In this book the word 'strategy' refers narrowly to conflict at the national level. During the Great War the term 'strategical' might have meant what we call 'operational' today. We also now see strategy largely as an issue of ways, ends and means: hence the structure of this chapter. That view considerably postdates the Great War. So we are not necessarily looking at the war as it was seen at the time.

To begin with ends and ways: the outline of Germany's 'Schlieffen' plan is well known.[1] The intention was to defeat France first and then turn against Russia. Germany attacked France with a strategic envelopment through Belgium, whilst defending in the east. However Germany's strategic *aims* at that stage are unclear. The plan was actually a blueprint for mobilisation, deployment, and operational manoeuvre. But why? What political ends did it meet? Naval plans included commerce raiding by cruisers and threatening British superiority in the North Sea. But, again, to what end?

1 Germany's plan is discussed in Chapter 3, but note the use of inverted commas here.

In 1914 the 'Schlieffen' plan failed. German strategy for 1915 focussed on defeating Russia in conjunction with Austria-Hungary. Significant gains were made, but Russia was not decisively defeated. Germany made very few attacks in the West that year. In 1916 Germany planned to break the French Army through a major operation at Verdun. That failed. Instead the German Army was ground down by a British and French offensive on the Somme. In late 1916 and early 1917 senior German commanders still believed that Germany could win the war. Militarily, Germany was in a highly favourable situation. Its armies were everywhere camped in enemy territory. It sought a compromise peace, but that was rejected.

But, strangely, the German Army apparently had no plan for major offensive operations in 1917. That may be true, but is grossly misleading. How did German commanders expect to win if they were not going to attack? To be fair, the evidence is not conclusive. It seems that Germany provided massive, covert financial support to Lenin's Bolsheviks. That was intended to convert the Russian February Revolution (which did not take Russia out of the war) into something like the October Revolution (which did). Meanwhile, Germany's resumption of unrestricted submarine warfare might cripple Britain (it did not). It might bring America into the war (it did). However, knocking Russia out of the war would enable Germany to move forces west for a war-winning offensive in early 1918, before the American Army could arrive in force. That offensive opened, against the British, on 21 March 1918. The overall offensive is generally called the *Kaiserschlacht* ('Kaiser's Offensive') and the initial battle 'Operation Michael'. It failed.

Turning to Austria-Hungary: Emperor Franz Josef was quite clear that he started the Great War. He did so to punish Serbia for assassinating his nephew, Franz Ferdinand. It would have been politically very difficult for Serbia to accept Austria-Hungary's ultimatum. However, for Austria-Hungary war against Serbia also meant war against Russia. Austria-Hungary therefore had to defend its northern territories (principally Galicia, in modern Poland) whilst invading Serbia. Austro-Hungarian campaigns against Serbia in 1914 and 1915 failed. Heavy casualties were suffered fighting Russia.

In Autumn 1914 Austria-Hungary had been dismayed to find that Germany had no immediate plans to invade Russia. So strategic planning had been flawed, and Austria-Hungary was at a strategic disadvantage from the beginning of the war. In the summer of 1915, Italy attacked

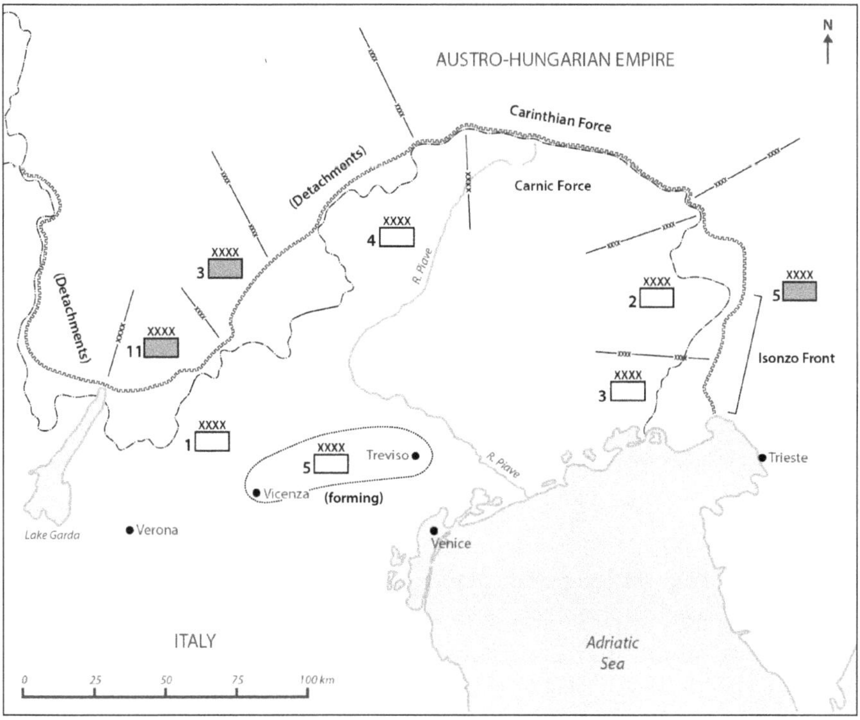

Figure 2-1. Northeast Italy in 1916

Austrian provinces in the Balkans (across the valley of the River Isonzo). Austria-Hungary now faced war on three fronts. Germany did support Austro-Hungary on the eastern front.

In Summer 1916 Austria-Hungary attacked Italy from the north through the South Tyrol. Some initial gains were made. The offensive then lost momentum. Russian attacks in Galicia forced Austria-Hungary to break off the operation and transfer forces north. Once again, Germany declined to support Austria-Hungary directly.

In 1917 Germany did provide some support to Austria-Hungary. The resulting counteroffensive at Caporetto (on the Isonzo) completely broke through the Italian defences. It resulted in a major advance as far as the River Piave. However, another Austro-Hungarian offensive in 1918, on the Piave, failed. An Italian, French and British counteroffensive cleared the Austro-Hungarian forces out of Italy in October.

France had clear reasons to go to war in 1914: not least, to throw invading German armies out of France and recover Alsace and Lorraine. Initial French attacks in 1914 failed. However, as we shall see in Chapter 3, France's defeat of the 'Schlieffen' offensive was well-handled and successful.

Nevertheless, Germany had overrun about a third of France's territory and much of its industry. In 1915 France's counteroffensives failed. It was forced to defend at Verdun in 1916, taking grievous losses in the process. Nevertheless, France made a significant contribution to the Somme offensive later that year. In 1917 the French 'Nivelle' offensive was a catastrophic failure. 49 divisions mutinied. By late summer only Britain and Italy were capable of major operations against Germany (described later). However, after Caporetto the Italians could not do so either. The French Army recovered by the end of the year. It played a major part in the defeat of Germany in 1918.

France also conducted major operations alongside Britain in the Dardanelles and Macedonia, also described below. The French navy played a significant part in containing Austro-Hungarian naval forces in the Adriatic.

Russia's reasons for entering the war in August 1914 illustrate classic Great Power diplomacy. Russia saw itself as the protector of the (south) Slavs. It declared war on Austria-Hungary in order to protect Serbia. Russia's standing in the Balkans had been eroded during the Balkan Wars, and a Serbian defeat would reflect badly on Russia. However, Russian mobilisation plans were crude. Mobilising against Austria-Hungary also meant mobilising against Germany. That prompted Germany to reciprocate. Furthermore, Russia had long-term objectives regarding Turkey, which would probably be drawn into the war. Additionally, Russia's economy would suffer grievously if the Bosporus was closed again, as it had been during the First Balkan War.

Russian armies had some initial successes against Austria-Hungary in 1914. They failed spectacularly against the Germans at Tannenburg. Turkey then attacked in the Caucasus. An early setback there led Russia to ask Britain to attack in the Dardanelles, in order to relieve pressure on the Russian Army.

1915 saw much fighting in Poland, with considerable Russian losses. The Brusilov offensive of 1916 brought some success against

Austria-Hungary. However, high casualties and dissatisfaction with the conduct of the war provoked a revolution in February 1917. Russian armies did attack in 1917 (the 'Kerensky Offensive') and continued to fight until the October (Bolshevik) Revolution. The Bolsheviks then accepted the humiliating Treaty of Brest Litovsk, which released 42 German divisions to fight in the West in the Spring of 1918.

Perhaps the most important aspect of Britain's strategy was its naval operations. The Royal Navy protected British maritime trade and Britain's links to its colonies. It conducted a four-year blockade of the North Sea. That tied up or confiscated 3.3 million tons of German shipping, leaving just 2.16 million tons available to Germany and Austria-Hungary combined. Germany could no longer trade outside Europe. That, in turn, contributed to gradual starvation. By early 1918 there was widespread unrest in Germany, partly due to food shortages.

Between 1792 and 1812 France (and Napoleon) had resisted the combined efforts of five major coalitions for 30 years. On each occasion one country in the enemy coalition 'broke ranks' and made peace. In 1914 Britain was determined that that would not happen again. The London Agreement of 5 September bound the members of the Entente (France, Russia and Britain) to seek no separate peace with members of the Central Powers. New allies (such as Romania and Italy) were obliged to make a similar undertaking. The concept of 'no separate peace' became very important. For example, Bolshevik Russia would later be excluded from the Paris Peace Conference because it signed the Treaty of Brest Litovsk.

In August 1914 Britain deployed most of its small Regular army to France; but to what end? Although never really enunciated, it seemed to wish to prevent Germany defeating France and hence becoming the dominant power in Europe. After a brave stand at Mons, the British Expeditionary Force (BEF) was forced to withdraw on the left wing of the French Army. It cooperated with the French counteroffensive, and so found itself in (French) Picardy and (Belgian) Flanders for the rest of the war. The BEF spent most of 1915 enlarging and conducting local offensives, broadly coordinated with the French. It grew from five to more than 60 divisions in time for a major offensive on the Somme in the summer of 1916.

German submarines constituted a real threat to British trade and hence its ability to continue fighting. Between February and June 1917 U-boats based in the Belgian ports (Zeebrugge, Ostend and Bruges (sic)) accounted for between 29 and 40 per cent of British monthly shipping losses. Britain's defensive minefield barrier across the Channel, maintained by the Royal Navy's Dover Patrol, was not working. Neither the Navy nor British Prime Minister David Lloyd George could see an answer. The War Policy Committee therefore endorsed a land offensive by the BEF to clear the Channel coast. That would also deprive the German air force of bases from which to bomb London.

In early 1917 the BEF fought at Arras in conjunction with the Nivelle offensive. It then shifted its efforts north, to conduct operations northward from Ypres. A fairly short advance would disrupt German rail access to Zeebrugge and Ostend, which supplied the U-boats. The battle of Messines was a precursor to what became known as the Third Battle of Ypres, or Passchendaele, that autumn. In 1918 the BEF successfully repulsed major attacks during the German Spring Offensive. It then led the counteroffensive that forced Germany to sue for peace in November. That counteroffensive began with the British attack at Amiens on 8 August.

Thus (perhaps by default) Britain's main effort was to defeat the main enemy, the German Army, in the field on the Western Front. Lloyd George repeatedly tried and failed to find an alternative approach. In late 1917 Britain decided to rebuild its Egyptian Expeditionary Force (EEF) to operate against, and potentially dismember, the Ottoman Empire. To do so it moved many of the EEF's British battalions, and any Regular cavalry, back to Europe. It doubled (that is, split and enlarged) many of the Indian units. Britain also seems to have decided, in the early summer of 1918, to seek the breakup of the Austro-Hungarian empire; although there was little that it could do directly towards that goal at first.

Partly at Russia's request, Britain had attempted to force the Dardanelles and then to invade the Gallipoli peninsula in early 1915. That would open the entrance to the Black Sea and force Turkey out of the war. Offering Constantinople to the Russians was probably the cheapest way to ensure that Russia would stay in the war. The Dardanelles operation probably did not shorten the war by a single day. If it kept Russia in the war, however, it was certainly worth attempting.

Having failed in the Dardanelles, Britain shifted its forces to support Serbia in a joint operation in (Greek) Macedonia in 1915. That campaign is generally known as 'Salonika' to the British. French, British, Greek and Serbian forces held back Bulgarian, Austro-Hungarian and German attacks until mid-1918. A French-led offensive then broke out of the perimeter at Salonika, French and Serbian forces marched north to liberate Belgrade. British forces entered Istanbul.

After initial reverses, by late 1917 British forces from India had secured oilfields in Mesopotamia (modern Iraq) in Ottoman Turkey. British forces in Egypt successfully resisted Turkish attacks against the Suez Canal before attacking into Palestine in 1917. That December the EEF entered Jerusalem. In September 1918 it broke through Turkish positions near Megiddo. British cavalry then advanced 80 miles in 34 hours. Its horses, and men, must have been utterly exhausted. The British pursued as far as the borders of Turkey, 340 miles from its start line, in a month. The Turkish Army in the Middle East was destroyed. Turkey lost its Arab provinces.

In November 1914 Turkey had been given financial incentives to enter the war on the side of the Central Powers. It wanted to recover some of the territory and prestige which it had lost in the Balkan Wars. It wished to profit at the expense of Russia, which it had been fighting for centuries. It could also benefit by seizing control of the Suez Canal from Britain. Turkey's operations in the Caucasus, the Dardanelles, Mesopotamia and Palestine were described above. The British failure in the Dardanelles freed up 13 Turkish divisions to operate elsewhere; particularly the Caucasus. When Russia withdrew from the Caucasus in late 1917, Turkey rapidly seized territory and oilfields in the Caspian region. Overall, Turkey spent much of the war shuttling forces between three or four fronts. Eventually it lost its empire and over 800,000 dead.

There was some coordination between Allied theatres and hence campaigns. It was generally limited to coordinating the dates of offensives. (Haig noted that the French Army was generally late in starting. There was often a delay of a week or so.) The movement of 11 French and British divisions to Italy in late 1917 helped keep Italy in the war.

Many Americans initially regarded the Great War as a European conflict which affected them little, if at all. America initially remained neutral. That was despite German U-boat attacks on American shipping and ships carrying American citizens (such as the British RMS *Lusitania*). German political machinations (such as the infamous Zimmerman Telegram to Mexico), subversion, and sabotage on American soil tilted American opinion towards war. The sabotage caused damage valued at $150 million; over $1billion today. The second German declaration of unrestricted submarine warfare, in February 1917, tipped the balance.

American war aims were vague until President Woodrow Wilson published his famous '14 Points' in January 1918. The overall intention seems to have been to impose a better peace on a belligerent Germany. 'Better' included, for example, abolishing European Great Power politics. That would, perhaps, give America more global influence. After declaring war, much of the US Navy was sent across the Atlantic to operate alongside the Royal Navy. America conducted a massive mobilisation in order to create an army to fight in France. The first troops fought in July 1918. 30 divisions saw combat.

In 1914 Italy was a member of the 1882 Triple Alliance with Germany and Austria-Hungary. However, by 1914 much of its economy depended on Mediterranean coastal shipping. Italy would have come off very badly in a naval war against France and Britain. It sought territorial gains, including in the South Tyrol and the Balkans, to which Austria-Hungary was unlikely to agree. Thus, after some thought, Italy entered the war against Germany and Austria-Hungary in May 1915. Limited advances northwards into the Tyrol were quickly halted. Italy's main effort focussed on eleven attacks across the Isonzo between 1915 and September 1917. After the disaster at Caporetto (the 12th battle of the Isonzo) in October, the Italians accepted French and British help to defend on the line of the River Piave (from east of Venice towards the northwest). Six French and five British divisions were dispatched. The final Allied counteroffensive, known as the battle of Vittorio Veneto, resulted in an Austro-Hungarian withdrawal in late October 1918.

Like Turkey and Romania, Bulgaria accepted considerable subsidies to enter the War. Bulgaria's war aims included reversing some of the results of the Second Balkan War, particularly in relation to Serbia. Bulgaria's entry into the war in October 1915 exposed Serbia's eastern border. An initial German and Austro-Hungarian attack from the north made some progress. A Bulgarian attack from the east then caused the Serbians to retreat through Montenegro to Albania. The Allies moved the remnants of the Serbian army by ship to join Franco-British forces in Salonika. German and Austro-Hungarian forces were gradually withdrawn. Bulgarian forces had also made a significant contribution to the defeat of Romania.

In 1918 Bulgaria asked Germany to stop withdrawing troops from Macedonia. Germany suggested that Bulgaria withdraw some of its forces occupying the Dobruja (in Romania). But why should Bulgaria choose to protect Macedonia (which it did not necessarily want), rather than Dobruja (which it did)? Germany displayed little interest in Bulgarian war aims. By late 1918 only a much-weakened Bulgarian army was left facing Allied forces at Salonika.

Romania's strategic interests included being able to sell its oil (the Ploieşti oilfields were the largest in Europe); sell its harvests; and gain Transylvania (where the largest ethnic faction was Romanian) and the southern Dobruja (where it was not). After considerable negotiation, Romania entered the war on the side of the Western Allies in August 1916.

Romanian operations in Transylvania initially went well. They were halted when forces were shifted south to the Danube (against Bulgaria). They then failed when German troops supported an Austro-Hungarian counterattack. In the south, the Romanians suffered a major defeat at Flamânda. That was followed by a successful German and Bulgarian advance into the Dobruja and then an inspired envelopment of the Romanian capital, Bucureşti. The remnants of the Romanian Army had to capitulate when the Russian army withdrew after Brest-Litovsk.

In the second half of the 19th century Japan had transformed dramatically into a modern, industrialized, empire-building and militarily aggressive nation. Its entry into the Great War in August 1914 reflected that. Japan sought territorial gains from German colonies in the Pacific or on the Chinese mainland. It gained them with little fighting and in cooperation with British and other western forces. Japan also deployed warships into the Mediterranean to support the Royal Navy.

So much for the ends sought, and the ways undertaken. What of the means applied?

In terms of navies, some new battleships were completed. The Royal Navy received 13; the USA and Germany six each. Cruisers and smaller vessels were built in large numbers, not least as convoy escorts. The British Royal Navy, for example, employed up to 193 escorts. Typically 125 were at sea at any one time. A few warships were built, or converted, to operate aircraft. There was more maritime shipping in commission, worldwide, at the end of the war than at the beginning: 45.7 million tons against 43.1 million.

Over 25,000 aircraft were built, but losses were very high. Most losses were due to accidents, often during pilot training. The British Royal

Air Force (RAF) was created on 1 April 1918 by combining army and navy air services. At the end of the war the RAF had about 2,000 aircraft in operational service. It had 200 squadrons at first line. 100 were on the Western Front and 63 were employed for home defence.

The best indicator of effort, however, was the number of army divisions raised. See Figure 2-2.

Nation	Total Raised	Deployed:								
		Western Front	Eastern Front	Balkans	Italy	Romania	Dardanelles	Egypt & Palestine	Mesopotamia	Caucasus
Germany	252	204	85	2		12				
Austria-Hungary	88		66	21	54					
Turkey	67 (2)			1		1	15	9	6	12
Bulgaria	16			14		7				
Total Central Powers	423	204	151	38	54	20	15	9	6	12
Russia	296 (3)		170			3				15
Romania	27					27				
Serbia	21			14						
Greece	11			11						
Italy	73	2		6	68					
France	152	115		9	5		2			
Portugal	2	2								
Belgium	8	8								
Great Britain:										
- UK	73	55		4	5		10	6	1	
- Australia	6	5					~2	2		
- Canada	4	4								
-India	13	5						2	9	
- New Zealand	1	1								
US	43	30								
Total Entente	706	239	170	33	78	30	14	10	10	15

Figure 2-2. Divisions Raised and Deployed

Notes (1): Includes Salonika.

(2): Several Turkish divisions were effectively destroyed and replaced. Hence the discrepancy in totals.

(3): Of which, 53 cavalry divisions.

The table is approximate. It generally shows the largest number of divisions in a given theatre: not the average, nor a typical number. Some divisions were destroyed; some of those were reformed. The German and Austro-Hungarian Armies, in particular, moved large forces between theatres.

A broader measure of effort by theatre and nation is given in Figure 2-3. It shows forces in terms of armies, or equivalents, in 1916 or 1917. Typically 9-12 divisions constituted an army, but not all forces of that size were called 'armies'. German corps were typically very large, so the number of armies understates the number of divisions. Forces of a corps or so, or less, are not shown. Figure 2-3 is intended to be illustrative only.

Nation	Western Front	Eastern Front	Balkans	Italy	Romania	Egypt & Palestine	Mesopotamia	Caucasus
US	2							
Germany	8	2						
Austria-Hungary		4		2	3			
Turkey						1	2/3	2
Bulgaria			2					
Russia		17			1			2
Romania					4			
Serbia			3					
Greece			1					
Italy			1/2	9				
France	9		1					
Britain	5		2/3			1	1	

Figure 2-3. Armies Deployed

Thus, for example, France committed the great majority of its forces to the Western front. Its other significant contribution was to Salonika. By contrast, only about two thirds of British forces were on the Western Front. After Gallipoli, the remaining third was divided between Palestine, Salonika and Mesopotamia.

Losses were horrendous. Figure 2-4 shows casualty figures for some nations.

Nation	Population (millions)	Manpower Served in Forces (millions)	Killed and Missing	Wounded	Civilian Deaths
Australia	4.87	0.42	53,560	155,130	-
Austria-Hungary	49.90	7.80	1,016,200	Up to 3,600,000	?
Belgium	7.52	0.27	38,170	44,690	c30,000
Britain	46.60	5.70	702,410	1,662,625	1,386 (1)
Bulgaria	5.50	1.20	77,450	152,400	c275,000
Canada	7.40	0.62	58,990	149,710	-
France	39.60	8.66	1,385,300	4,239,200	c40,000
Germany	67.00	13.40	2,037,000	5,687,000	c700,000
India	316.00	1.68	62,060	66,690 (?)	-
Italy	35.00	5.90	462,400	955,000	?
Japan	67.20	0.80	1,970		
New Zealand	1.05	0.13	16,710	41,320	-
Romania	7.51	?	219,800	?	>265,000
Russia	167.00	12.00	c1,800,000	c4,950,000	c2,000,000
Serbia	5.00	0.71	127,500	133,150	c600,000

Figure 2-4. Casualties
Note: (1) In air raids. Some others in coastal shelling by the German navy.

Very large numbers were deployed, killed, and wounded. Few countries suffered many civilian deaths due to enemy action, but some suffered badly through starvation and disease. The great majority of military casualties were soldiers. Few were sailors and very few were airmen.

It was not 'the death of a generation'. However, losses could be extremely localised. For example, many men from a given village or town, or almost all of a class of an Oxford or Cambridge college, might have been killed. Thus the effect on public opinion or decision makers was often marked.

Germany was beaten. It was beaten on land, on the Western Front. On 11 August 1918, three days after the British attack at Amiens, the German chief of the General Staff (Erich Ludendorff) said that 'this war must be ended'. That was the background to Allied armistices with Bulgaria on 29 September, Turkey on 30 October and Austria-Hungary on 3 November. Germany signed an armistice on 11 November.

There were clear winners and losers. Four empires ceased to exist. Germany lost part of what became Poland, Alsace and Lorraine, and all of its overseas possessions. Turkey lost its Arab provinces. Austria-Hungary was broken up into two core territories (Austria and Hungary). Its other provinces formed new independent states. Russia lost Finland, the Baltic states, Ukraine (briefly) and part of Poland.

The winners included several new European states. France, Japan, Britain, New Zealand, Australia and South Africa gained territories. (With the exception of German South West Africa to South Africa, all were overseas.) However, any gain from 'mandated' territories outside Europe was typically not great. Within 30 years most of the former Ottoman Arab provinces became independent.

The German Navy – the main threat to Britain – disappeared. Germany was disarmed for over a decade, benefitting Britain and France. America and Canada gained relatively little. A new world order was promised with the establishment of the League of Nations but, as we shall see, that promise was not delivered.

The main problem with the outcomes of the Great War is that, to many people, the benefits were not obvious. For example, Britain gained a virtually guaranteed source of oil. Before the war that was not seen to be important. After the war it was taken for granted. Yet the losses – the costs – were seen as grievous.

What can we observe or deduce from looking at the Great War at the strategic level? Firstly, the terminology is unhelpful. For example, we tend to talk about the *Battle* of the Somme and the Third *Battle* of Ypres. Yet those were the main operations undertaken by the BEF in 1916 and 1917 respectively. We can identify 12 named 'Battles' on the Somme in 1916 (the Battles of Albert, Bazentin, Delville Wood and so on), and similarly eight around Ypres in 1917. In both cases the area of operations was quite small. However, we can understand the Great War better if (in these and similar cases) we refer to the Somme *Campaign* of 1916 and the Passchendaele *Campaign* of 1917.

This is not just terminology. Strategy is done by planning, resourcing and conducting campaigns in theatres of war. Much of the Great War was waged by conducting protracted, major operations on the Western Front. Those operations typically had objectives at the national and alliance level. There were broadly coordinated with other fronts and nations. We should see the conduct of war on the Western Front as a series of discrete campaigns. They were not just a series of seemingly pointless, bloody slogging matches.

Secondly, it was entirely reasonable for Britain to decide to attack Turkey via the Dardenelles. As the British Prime Minister Herbert Asquith remarked, the potential (strategic) gains were enormous. In practice, Britain gave away strategic and operational surprise. It planned a purely naval operation which was so flawed that the admiral chosen to command it (Admiral Sir Percy Scott) declined the appointment. Britain then undertook land and naval operations separately, ignoring the advice of the Director of Military Operations (Major General Charles Callwell), a coastal artillery expert who had reconnoitred the area before the war.

Nevertheless, the Dardanelles and Gallipoli represent a reasonable strategic idea executed poorly. More typically, nations were good enough at the 'what' and 'how' of high-level planning (the ways and means), without adequate connection to the 'why' (the ends or goals). Mobilisation, deployment and operational plans (such as Germany's plan for 1914) are not, in themselves, strategy. Strategic aims may, naturally, change during a long war. It is, however, surely a failure of strategy if war plans do not consider what high-level goals are sought, and indicate how military operations will be used to achieve them.

The examples of (for example) Italy, Japan, Romania, Bulgaria and perhaps Turkey demonstrate very strongly the importance of self-interest. Germany seems to have assisted Austria-Hungary only when doing so suited German interests. War is the extension of politics using violent means. Here 'politics' implies the creation of relative national advantage. Logically, nations should not go to war unless they can identify some benefit (that is, some self-interest) in doing so. That advantage should be expressed as concrete goals. They are the ends sought. Such self-interest is not cynical. It should be the very essence of war.

There has been much talk of the 'Fourteen Points'. But look closely at the Armistice document imposed on Germany on 11 November 1918. It does not resemble the 14 Points at all closely. Then look at the five treaties signed by the losers (Germany, Austria, Hungary, Bulgaria and Turkey) after the Paris Peace Conference in 1919. Although they did collectively achieve some of the 14 Points, some others were already redundant (for example, Belgium was already liberated, meeting Point Seven). Other issues were well beyond the provisions of the 14 Points. They include so-called German 'war guilt' (discussed later) and the demilitarisation of the Rhineland.

The Peace Conference reshaped whole continents. It imposed settlements on the losers. Unsurprisingly that caused some resentment. But the key shortcoming of the Paris process was not in the shape or content of those settlements. It was that the resulting peace did not last.

No political issue is ever resolved for all time. But if major wars are to be fought for major political goals, then surely those goals should be identified clearly. The settlement reached should then be enduring. However, for all the fanfares at Paris, the supposedly victorious Allied powers failed to achieve a worthwhile, lasting outcome to the Great War.

This is *not* an issue of 'winning the war but losing the peace'. Winning the war *is* winning the peace. 'Winning the war' means gaining a better political outcome. The fighting is just the means of achieving that. 'Winning the fighting' is *not* winning the war.

On a different subject: in economics, costs or investments made in the past are foregone. When assessing courses of action, they should be disregarded. Future undertakings should be made only on the basis of assets and liabilities. That is fundamental. Failing to recognise that leads to 'doubling down' on past investments. It is the cause of many bad business decisions. That applies to national, political decision making as well as to business. That has a major implication for the way we see wars, and particularly the Great War.

In war, casualties are out*puts*. They are always tragic. They typically have (entirely understandable) emotional, social and political consequences. They are, nonetheless, outputs. At any point, when considering future warlike activity, past casualties should be seen as losses. That is, negative outputs: costs foregone. Conversely political change, and especially national political gains, are out*comes*.

The difference is critically important. Past casualties, and particularly lives lost, should not be the deciding factor in strategic decision making. Clearly commanders, and political decision makers, should seek to minimise losses; particularly in terms of casualties. But to compare political outcome with casualties suffered is to compare apples with pears. It is widespread. It is entirely understandable. Socially and politically, it will be tremendously difficult to avoid. It is a reasonable assessment *after the event*. But, when considering courses of action, it is flawed. After the event it is far more useful to compare relative political gains and losses, and how they endure.

Take, for example, the situation on the Western Front in the summer of 1918. France and Britain had suffered millions of casualties, including hundreds of thousands of dead each. But those casualties *had been* suffered, regardless of future courses of action. By way of assets France and Britain both had strong, capable and extremely experienced armies. They had an ally – America – whose armies would only grow stronger. They had an opponent whose army had failed in its last major offensive, and which had exhausted its last reserves of manpower. Its allies were falling apart.

By way of liabilities, Britain and France had the very real costs of maintaining forces in the field, and also any casualties which might occur. Another liability (in this case political) was the risk that America would demand an increasing say in any outcome as the war progressed. In practice, what appears to have happened is that casualties already suffered (that is, costs foregone) were allowed to outweigh most other considerations. That was described as 'war weariness'. It was entirely genuine. Critically, however, it was a social and *to that extent* political phenomenon. A more dispassionate calculation might have considered, for example, whether (say) a further quarter of a million Allied casualties would have brought about the occupation of Germany and the imposition of a more enduring political settlement (like that imposed after the Second World War).

We shall look at the historiography of the Great War in more detail in Chapter 3. Here we should remark that, for the British, much of our perspective arises from Prime Minister Lloyd George's memoirs, published between 1933 and 1936. However, Lloyd George was not a nice man. Both he and his memoirs can be seen as small-minded, self-serving and vindictive. And Lloyd George was, of course, one of the main architects of the Paris peace process.

3

The Great War: Campaigns and Operations

The operational level of war was not clearly identified during the Great War. Much of the written history still reflects that. So considering the events of 1914-18 from a campaign and a theatre perspective can provide fresh insights.

Chapter 3 looks at naval operations and then considers three aspects of theatre-level operations on land. They are: the so-called 'Schlieffen' Plan; the campaign in Romania in 1916; and Field Marshal Sir Douglas Haig's command of the British Expeditionary Force (BEF) in France and Flanders. The chapter then considers some high-level personnel and logistic issues and some aspects of the historiography.

We can describe major naval operations quite simply. In 1914 the Royal Navy cleared the high seas for allied use. It contained the German surface fleet in the North Sea. The Royal Navy, the French and subsequently the Italian and Japanese navies contained the Austro-Hungarian fleet in the Adriatic. That established a blockade in both cases. The Central Powers initiated submarine warfare on trade which, although problematic for the Allies, failed. Allied control of the high seas allowed, for example, millions of American soldiers to reach Europe safely. (Only 63 were lost at sea to enemy action.)

Within days of the outbreak of war, German cruisers were sinking British merchantmen around the world. They sank 203,000 tons of shipping in 1914. In turn those cruisers were all sunk by March 1915. The only German naval formation at sea, the Far East Squadron, was destroyed off the Falkland Islands in December 1914. The blockade of Germany remained in place throughout the war. By late 1917 food rationing in Germany was having a serious effect on public health, morale, and public sentiment.

In August 1914 two German warships, the *Goeben* and the *Breslau*, evaded Royal Navy forces in the Mediterranean and entered neutral

Turkish waters. Turkey had no Dreadnought battleships. Russia had none in the Black Sea. The *Goeben* was a Dreadnought battlecruiser (*Breslau* was a light cruiser). The two ships were effectively transferred to the Turkish navy and then operated in the Black Sea. Due to the absence of any other Dreadnoughts, effective naval aviation or hostile submarines, the *Goeben* dominated the war in the Black Sea. This is a rare but clear example of one-sided technical advantage at the operational level.

The most dramatic naval encounters of the Great War took place in the North Sea. They include, for example, the Battle of Jutland in mid-1916. But consider the issue from a theatre perspective. Almost the whole German surface fleet was bottled up in the North Sea for the duration of the war.

The Dardanelles were an effective barrier to submarines in both directions. Austro-Hungarian and subsequently German submarines operated from bases in the northern Adriatic. More than half of the merchant ships lost by the Allies were sunk in the Mediterranean. Many were sunk by German U-boats using Austro-Hungarian ports. Allied operations at Gallipoli and Salonika had to be conducted in the face of a credible submarine threat. Sea routes to and from the Far East via the Suez Canal were not significantly affected.

Turning to land campaigns, it is bizarre that the failure of the Schlieffen plan is so well known. For example, British high school pupils were taught about it for decades. The reason for that will be discussed later.

Schlieffen apparently intended that the German right wing, advancing through Belgium, would outnumber the left wing by eight to one. Germany had 85 divisions available. An eight-to-one ratio would mean about 76 divisions on the right and nine on the left. In 1914 the two right-wing armies possessed 27 divisions, but the road and rail networks through Belgium were strained to the limit. France had 81 divisions. The BEF had five.

The German plan required the right wing to reach a line from Abbeville through La Fère to Diedenhofen by 1 September. See Figure 3-1:

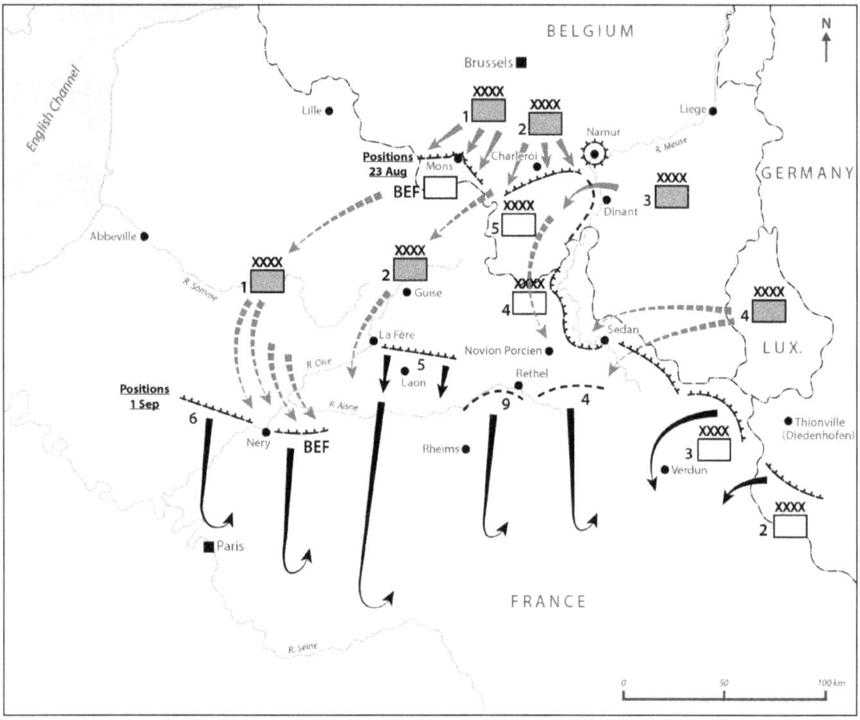

Figure 3-1: The Execution of the 'Schlieffen' Plan, 1914

On 1 September the right wing was, if anything, ahead of schedule. However, by 15 September the French Commander in Chief, Marshal Joseph Joffre, had created two new armies (the Sixth and the Ninth) on his left wing. He had committed 18 divisions from his reserves, nine from his right wing, and one from Morocco. The German attack was halted and turned back on the Marne. A race to the North Sea, to outflank the opposition soon followed. The opposing armies reached the English Channel in late October. The result was four years of trench warfare.

The German plan was not an act of genius. The French could, and did, shift reserves across interior lines faster than the Germans could march round through Belgium. No 'what might have beens' can disguise that. Such 'what might have beens', or 'what ifs', include Moltke the Younger *not* weakening the right flank; nor withdrawing divisions to reinforce Russia. Simply wheeling a massive force through Belgium would not, of itself, defeat the Allies. The stronger the German right wing, the more the French could reinforce their left wing. Put simply, the German plan largely ignored France's ability to react.

On two occasions (on 23-24 August at Dinant and on 28 August at Guise), German forces found gaps which forced the French to withdraw. Doing so effectively accelerated the German advance. On either occasion, significant French forces might have been encircled and destroyed. However the German plan relied on a relentless, theatre-wide encirclement which would not accommodate local variations. To repeat: the 'Schlieffen' Plan was not an act of genius. The key issue here is that discussion of the plan almost invariably considers German intentions (and actions) in isolation. Yet war is unavoidably at least two-sided, and adversarial.

To understand the events in Romania in 1916 we need to understand the geography. See Figure 3-2.

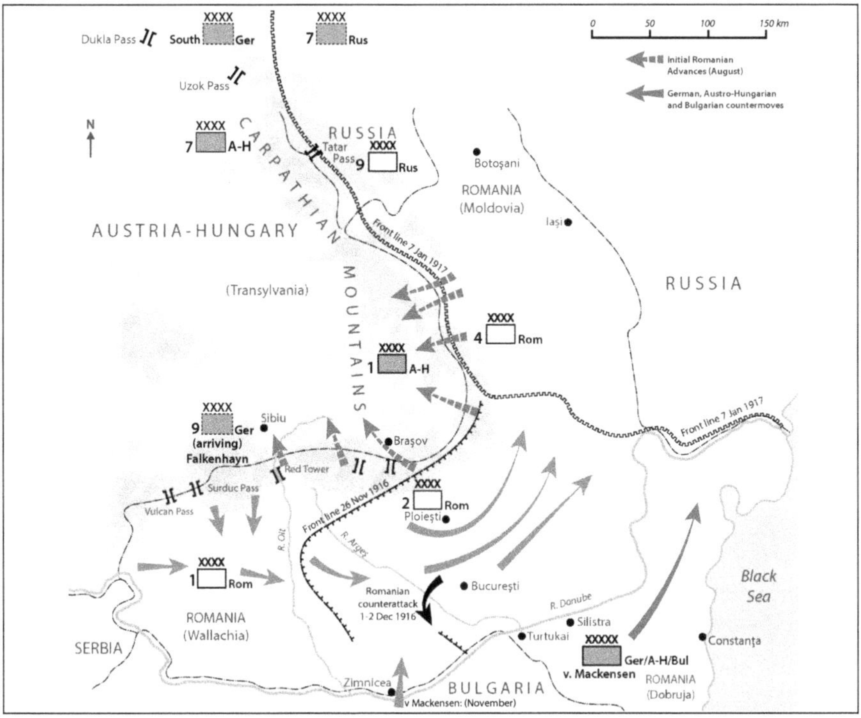

Figure 3-2: Romania in 1916

A rapid Romanian advance northwest along the western side of the Carpathians might have opened the Uzok and Dukla passes. Russian armies could then have flooded into Hungary and brought about the collapse of Austria-Hungary. General Erich von Ludendorff, at the time chief of staff to the German commander in chief in the east (Field Marshal Paul von Hindenburg), certainly thought that might happen.

Romanian aims, however, were limited to seizing Transylvania. Its armies were reasonably successful until halted by Austro-Hungarian and German forces. The Romanians were then forced back to defend the Carpathians, and particularly the southern passes. However Romania now faced a war on two fronts. The German field marshal August von Mackensen was sent to Bulgaria to command an *ad hoc* 'army group'. It consisted of four Bulgarian and one Turkish divisions, together with a few German battalions. Von Mackensen advanced into Dobruja. Once reinforced by a single German division, he broke through a Romanian fortified position from Rasova to Tuzla.

That regained most of Dobruja for Bulgaria. It also denied Romania the use of Constanţa, its only Black Sea port. In an inspired move, von Mackensen then moved west along the south bank of the Danube and crossed at Ziminicea on 23 November. At about the same time the German Ninth Army forced the Vulcan and Surduc passes. A motorised force then opened the Iron Gates (along the Danube into Serbia) from the east.

The Romanian First Army, defending western Wallachia, was at risk of being encircled. It withdrew eastwards towards Bucureşti. As von Mackensen's forces advanced to cross the river Argeş, they were attacked in the flank by a sortie from the Romanian capital. A cavalry division from the German Ninth Army linked up with von Mackensen. The Romanian attack was defeated, and Bucureşti fell on 6 December. The Romanian First Army got away to join the remaining Romanian forces near the Russian border.

It had been a four-month campaign of manoeuvre. 25 Romanian divisions had fought along a 1,600-kilometre front. There had been heavy fighting in the Carpathians. Von Mackensen's switch from Dobruja to Ziminicea involved marching over 100 miles west, but threatened to encircle the Romanian First Army. The First Army successfully withdrew over 350 miles. Cavalry forces had played an operational, that is theatre-level, role. We tend to think of the Great War as an unrelating, attritional slog of trench warfare. Much of it was, particularly on the Western Front; but by no means all of it.

When considering Haig as a commander we should note that no general, of any army, achieved a tactical breakthrough on the Western Front during the Great War. Haig took over command of the BEF from Field Marshal Sir John French in late 1915. In effect he commanded four major operations: on the Somme in 1916; at Ypres in 1917; defending against the German spring offensive in 1918; and in the 100 Days' offensive in late 1918. The battles of Arras in early 1917 and Cambrai later the same year are also operationally significant.

The war on the Western front is often understood as long periods of routine and boredom in the trenches, interspersed with the occasional major battle. From the perspective of a given battalion, that may be true. However, consider the picture from the operational level. From the beginning of the Somme offensive to the end of Passchendaele was a period of 16 months. In all that time the BEF attacked continuously (on the Somme, at Arras, Messines, Passchendaele and Cambrai) for all but sixteen weeks.

There was then a pause as the western Allies awaited the *Kaiserschlacht*. The German offensive ran almost continuously from 21 March to 18 July 1918. The Allied counteroffensive opened three weeks later (on 8 August) and ran continuously to the end of the war. So, with the exception of the period when the German Army had the initiative, the Allies (and particularly the BEF) were attacking, somewhere, more than 80 per cent of the time.

In early 1916 the BEF reached its full strength. 63 British, four Canadian, five Australian, one New Zealand, five Indian and two Portuguese divisions fought with the BEF on the Western Front. Three of the five armies, 11 corps and almost every British division fought on the Somme.[1]

Haig attacked on the Somme on 1 July 1916. He had three objectives. None related to terrain features. All had strategic significance. They were: to relieve pressure on the French at Verdun; to prevent the movement of German forces to other theatres; and to contribute to wearing out the German Army to the point where it could be beaten. Haig was successful

---

1 One division, the 61st, did not fight *on the Somme*. It attacked at Fromelles in a supporting operation on 19 July and took significant casualties. It had only arrived in the BEF in late May (1916). That apart, *every* division serving in the BEF at the time fought on the Somme.

to some extent in all three objectives; but none of them completely. German attacks on Verdun stopped on 11 July. Six divisions *were* transferred from the Somme to other theatres. However, at the beginning of the campaign there were 107 German divisions on the Western Front. At the end there were 117. Thus there was a net *inflow* of 10 divisions during the campaign.

97 German divisions served on the Somme. The best estimate is that they suffered an average of 4,479 casualties per division. 29 divisions served twice. Three divisions served three times. That equates to a total of 132 rotations into the front line. Thus the average loss was 3,291 casualties per division, per rotation. German divisions had a nominal strength of about 13,000 men. Over 3,000 (let alone 4,000) casualties per division represents a significant 'wearing out', particularly among the infantry. On 5 November Haig received a captured German document, dated 14 October. It stated that a sample of 528 German battalions (about half of all those which fought on the Somme) had already suffered 46 per cent of their established strength as casualties. Operations on the Somme involved 83 per cent of all the German divisions in the west. The German experience on the Somme has been described as 'the death of the prewar German Army' and 'the bloody grave of the German field army' by German officers.

For the British the Somme was a linked series of twelve named battles. In modern terminology that is a campaign. Haig explained his rationale, his objectives and the events of the campaign, in a dispatch published soon after.[2] Several writers who have criticised the battle and Haig's conduct of it do not seem to have read that dispatch. To repeat: none of his objectives related to terrain. None required a tactical breakthrough.

As a result of the Somme campaign, the Germans conducted their only operational withdrawal on the Western Front between Christmas 1914 and the summer of 1918. That withdrawal, in early 1917 (to the Siegfried Line), shortened the German front line. That freed up 14 divisions for use as reserves. However, it also shortened the Allies' front line, with much the same effect.

The withdrawal pre-empted the Allies' Spring offensive for 1917. The Germans had laid waste to the area over which they withdrew. It would not support Allied offensive operations. Instead of attacking alongside each

2 General Sir Douglas Haig, *Dispatch*, 23 December 1916. Published as a Supplement to the The London Gazette on 29 December 1916.

other, the French would have to attack further south, along the Chemin des Dames. Haig had to attack further north, at Arras. The French 'Nivelle' offensive was both late and a bloody failure. British attacks at Arras were generally as successful as those of the later stages of the Somme campaign. The Canadian seizure of Vimy Ridge, substantially on the first day, was the high point both literally and metaphorically. Significantly, Canadian and British forces captured large numbers of German prisoners for the first time. The failure of the Nivelle offensive made the Arras attack operationally pointless. It was closed down.

As mentioned in Chapter 2, 49 French divisions mutinied as a result of the Nivelle offensive. The February Revolution meant that Russia could not be relied upon to fight. Only Britain and Italy could now actively engage the Central Powers. As Arras drew to a close, Haig prepared to clear the Belgian coast by attacking north from Ypres. Zeebrugge was close to the border of the neutral Netherlands. An advance of just a few kilometres from Ypres could either cut off or lead to the capture of Zeebrugge.

The Messines ridge overlooked the staging area for the Passchendaele campaign. So the battle of Messines can be seen as a preliminary operation for Passchendaele. The Germans had held the ridge since 1915. British engineers had been preparing to blow the top off the ridge, almost literally, for months. On the morning of 7 June 1917, 20 mines (containing over 400 tons of explosive) exploded under the German defences. To ensure surprise, there was no preliminary bombardment. The noise of the explosion was heard in London, 130 miles away. The British attack was a complete (although costly) success: it seized the operational objective.

Like the Somme campaign, Passchendaele was a linked series of battles. Each had a distinct tactical objective, which was typically a given section of the German defensive position. Army commanders planned a succession of operations which were typically executed as corps-level attacks. Stubborn German opposition and foul weather severely limited progress. The capture of the Passchendaele ridge was meant to be followed by an advance to the railway at Roulers (Roeselare in Flemish). Passchendaele village was captured on 6 November. Even worse weather, and the second (Bolshevik) revolution in Russia on 7 November, led to the end of the Passchendaele campaign.

We can now identify what we might call Haig's 'operational method'. He moved the HQ of Gough's Fifth (formerly 'Reserve') Army into the line both on the Somme and at Passschendaele. Thus there were two Army HQs commanding operations alongside each other in both cases. Corps HQs were largely kept in the line. Divisions were rotated through Corps. Thus, for example, 25 divisions served under HQ X Corps on the Somme.

On 20 November two corps of the British Third Army attacked at Cambrai, about 60 miles south of Ypres. It was completely unexpected. The British used a very short preliminary bombardment and 437 tanks. It was the first genuinely massed tank attack ever. It achieved complete operational surprise. The British advanced about four miles, much of it on the first day. The Germans rushed reinforcements to the area. Within a few days they recaptured most of the ground they had lost. The British learnt two major lessons. The first was that large-scale attacks could achieve operational surprise and penetrate well into the depth of German defensive positions. The second was that, critically, British divisions did not know how to consolidate and defend the ground they had taken.

The Bolshevik revolution and the Treaty of Brest Litovsk allowed Germany to transfer about 40 German divisions to the Western Front by the spring of 1918. The Allies prepared for a major German offensive which would take place before significant American forces could arrive later that year. British divisions were reorganised (as discussed below) and trained in new defensive tactics. Those tactics incorporated several of the lessons of Cambrai.

The *Kaiserschlacht* was intended to split the BEF from the French Army and force the British back to the sea. That was the *intention*, but the initial attack (Operation *Michael*, against the BEF on 21 March 1918) had no operational *objective*. The Germans benefitted from novel 'Stormtrooper' tactics, a hurricane artillery barrage, the generous use of gas, and a very foggy morning. The Germans attack reached the back of the British Battle Zone by the end of the first day. The British Rear Zone was not penetrated. The Army commander, General Hubert Gough, *ordered a withdrawal* across the Crozat Canal that night. By doing so he pre-empted the German attack the next day to some extent. The British Fifth Army then conducted a fighting withdrawal as French and British reserves were brought up. Gough narrowly managed to maintain a coherent delaying operation. The German advance was halted just east of Amiens, about 40 miles from its start line. A secondary German attack, on the Third Army on 28 March, was repulsed easily.

There were five further German attacks on the French and British armies. They were all contained after several days' fighting and some loss of ground. Either the French and American attack on the Marne on 18 July, or the British attack at Amiens on 8 August, can be considered as the start of the Allied counteroffensive. The '100 Days' are considered to run from 8 August to 11 November 1918.

At Amiens the British Fourth Army conducted a surprise attack with four corps and 532 tanks. Once again, it was completely unexpected. It was the first time that an army-level operation achieved surprise on the Western Front. Ludendorff, now effectively chief of the German general staff, went into shock (albeit briefly). He called it 'the Black Day of the German Army'. Three days later he said that 'this war must be ended'. At Amiens the BEF penetrated the whole depth of the German position on the first day. It failed to break out, due to poor coordination with the cavalry. There were no German reserves in the area that day. That was a direct consequence of operational-level surprise. German divisions were rushed up. However, tellingly, their counterattacks failed.

Amiens was the first of a linked series of Allied offensives. None broke through the German defences. All pushed the German armies back, took considerable prisoners, and seriously weakened the Germans. By November German battalions typically consisted of little more than a few officers, a few soldiers and a few machineguns. They lost ground almost every time they were attacked. In a sample of about 100 British divisional-level attacks from this period, 63 per cent gained their objectives as planned.

In the 100 Days the western Allies employed a sophisticated operational approach. They were coordinated by the French General Ferdinand Foch. Typically, an Allied army would attack and achieve some success. It would draw in German reserve divisions, and then close down the offensive after a few days. As that attack was coming to an end another army would attack elsewhere; and so on. The Germans ran out of reserve divisions. Increasingly larger numbers of Germans surrendered. In the British sector much more territory was abandoned by the Germans than was seized by the BEF. That was because successive BEF attacks outflanked German defensive sectors, forcing the Germans to withdraw. French and American attacks had much the same effect.

As discussed later, the German Army *was* defeated in the field on the Western Front in 1918. The BEF advanced further than any other

Allied army: about 100 miles. It captured more prisoners than the French, American and Belgian armies combined. The BEF also lost over 100,000 dead in the process. So, in simple terms: a hundred days, a hundred miles and a hundred thousand dead.

Given that the German Army *was* defeated in the field; that Haig was the British commander in chief; and that *no* army achieved a breakthrough on the Western Front, it is hard to criticize him. Haig was very popular with his subordinates and his troops. When he died in 1928 his body lay in state in both London and Edinburgh. In total, about a million people turned out to pay their respects. Objectively, it seems that habitual criticism of Haig is (and has long been) either ill-informed, nothing to do with his performance as a commander, or both.

In June 1914 the Serbs had selected, trained and equipped Franz Ferdinand's assassins and given them their orders. In 1917 Germany bankrolled Lenin and the October Revolution. It had funded sabotage in America. It supplied arms and ammunition to the Easter Rising in Ireland in 1916. Britain paid an Italian informer named Benito Mussolini. A British intelligence officer ('spy'?) destroyed the Ploieşti oil production facilities in Romania before the Germans overran them in 1916. Britain bribed the Arabs to revolt in the Near East with millions of pounds of gold in 1918. The spy novels of John Buchan (such as 'The Thirty-Nine Steps' and 'Greenmantle') now seem tame compared to those of Ian Fleming. Buchan worked in the British War Propaganda Bureau and then as an army intelligence officer. Fleming worked in naval intelligence in the Second World War. What made Buchan's novels more interesting at the time is that they came out *during* the Great War. James Bond did not appear until 1953. Clearly espionage, subversion and other irregular activities were fairly widespread in the Great War.

We should briefly look at high-level organisational, personnel and logistic issues. At the beginning of the war most armies had divisions organised as companies, battalions, regiments *and* brigades. The exceptions were the Turkish and British armies. During the war divisions were generally streamlined by removing either regimental or brigade headquarters. They were also reduced from 12-18 battalions to 9-12. Not least, that made formations quicker to respond to new orders. German advisers had reduced Turkish divisions to nine battalions before the War, and they kept that structure. That reduction reflected lessons from Russo-Japanese war.

We now know a lot about the officer structure of armies and how that changed through the war. In the British case, by 1918 battalion commanders were almost all under 35 years old. Some were as young as 24. Brigade, and some divisional, commanders were also much younger than in peacetime. That was not the case at corps and army level. There were some shockingly bad senior commanders. They included General Sir Hubert Gough, who had been promoted to command the Fifth Army whilst younger than all other *corps* commanders. He was sacked in the Spring of 1918. Another example is Field Marshal Sir Henry Wilson. Wilson never commanded a division nor an army and made a mess of commanding IV Corps in 1916. Yet he became Chief of the Imperial General Staff. Gough and Wilson were highly charismatic intriguers, and probably classic authoritarian personalities.

As early as the later stages of the Somme, the German Army was having morale problems. Nine of the divisions that fought on the Somme were described as 'unreliable'. By late 1917, tens of thousands of German soldiers had deserted to the Netherlands. Conversely, Haig remarked that during the withdrawals of March 1918 that morale in the BEF was as high as it had ever been.

Logistics were a monumental problem, especially in terms of artillery ammunition. The British Army expended 2.38 million tons of shells for the preliminary bombardment on the Somme alone. Expenditure during the 100 Days averaged 2.08 million tons *per week*. Yet there were no four-wheel-drive trucks to deliver the ammunition to the guns. They had not been invented.

Railways were central to effective logistics. In 1916 and 1917 the construction of light railways often indicated an impending offensive. They were easily detected on aerial photographs. The BEF's railway system almost broke down in 1916. The problem was solved by appointing a civilian to be director of transport as a major general. In practice, the biggest single difficulty was often to bring forward enough roadstone, by rail, in order to repair railways (or build new ones).

There were major improvements to the evacuation and treatment of the wounded. On the first day of the Battle of Arras, casualties were being offloaded from trains at Charing Cross station in London soon after noon. Zero Hour had been at 0530hrs.

Not all armies could learn as fast, nor at the same time. The US Army's transport system broke down in the Autumn of 1918. Thousands of soldiers went without food for days. American formations reported thousands of 'stragglers' missing from their units. This is not the picture of a bad army. It is a picture of a brave but massively enlarged and inexperienced army trying to learn very quickly.

The historiography of the Great War has much to answer for. The issue of the supposedly flawless 'Schlieffen Plan' was concocted by the German General Staff after the War. The purpose was to reinforce the image of the General Staff as a supremely professional institution which could have won the War in 1914 if it were not for Moltke the Younger. The German public, and Anglo-Saxon writers, swallowed it unthinkingly. The fact that Moltke was the then *Chief* of that General Staff was quietly ignored.

More recent historiography can be just as disappointing. Between 1999 and 2015 two well-known academic historians staged a protracted and at times bitter debate in print as to whether there ever was a 'Schlieffen Plan'. The conclusion seems to be that:

- there was no 'Schlieffen Plan' in the form of a General Staff order or directive.
- Schlieffen conceived of a 'strong right hook' through Belgium with two powerful armies.
- the troops notionally allocated to the plan by Schlieffen did not exist in reality.
- what Moltke the Younger did in 1914, with the troops actually available, did not meet the expectations of Schlieffen's concept.
- after the war, the German army establishment effectively wrote the official history so as to blame Moltke the Younger. It emphasized the apparent brilliance of Schlieffen's planning, and hence the professionalism of the General Staff.

The first, second and fourth points are no surprise at all. The third and fifth constitute a minor corrective. Yet the effort, the vitriol, and the more than 150,000 words published was staggering. Grown men published articles with titles such as 'Terence Holmes Reinvents the

Schlieffen Plan – Again'[3]. One editorial called it 'history at its best'.[4] Seen from outside the confines of academic history it seems trivial.

The German 'stab in the back' myth is just that: a myth. Apparently the German Army was never defeated in the field. Instead, it was betrayed by profiteers on the home front, and particularly Jewish financiers. The myth was useful to the German General Staff after the War. It was also very popular with sections of German public opinion. Conversely the German General Staff did not fool itself. It simply published, in its Official History, extracts from corps and army operational reports. They demonstrated, absolutely clearly, that the German Army was in fact defeated.

The history of the Great War is generally dominated by the history of the Western Front. That gave western readers, both military and civilian, a demonstrably biased understanding of the war as a whole. The Romanian campaign of 1916 included much long-distance, theatre-wide manoeuvre. So did several other campaigns in eastern Europe. So did Allenby's campaign in Palestine in 1918. The conceptual imbalance led western thinkers to focus on a future war like warfare on the Western Front. German thinkers thought otherwise.

Some of the early published histories were misleading. Volumes of Churchill's 'The World Crisis' appeared from 1923 onwards. Churchill wrote outstanding English, but he was no historian. As an example, his chapter on the Somme is stuffed with hyperbolic prose, and he was aware of Haig's Dispatch, but he either misunderstood it or chose to ignore it.[5]

Finally, we come to Haig. His memoirs were not published until the 1950s. Lloyd George's memoirs *were* published, over several years, in the 1930s. By then many other politicians and commanders had already published their memoirs. Several were highly critical of Lloyd George. He deflected that with a critique of the senior generals and, at least by implication, Haig. Lloyd George's agenda was twofold: self-promotion and destroying the reputation of the generals; particularly Robertson (the CIGS) and Haig.

---

3 Terence Zuber, 'Terence Holmes Reinvents the Schlieffen Plan – Again', *War in History*, 10:1 (January 2003), 92-101.

4 Hew Strachan and Denis Showalter, Editorial, *War in History*, 11:1 (January 2004), 1-2.

5 Winston Churchill, *The World Crisis 1911-1918*. (Abridged and Revised Edition. London: Penguin Books, 2007), 652-668.

Regrettably, he largely succeeded in the latter aim. To illustrate that: Lloyd George wrote over 100 pages on operations at Passchendaele, but only 28 to the considerably more successful events of the 100 Days. Lloyd George's, and Liddell Hart's, comments reinforced general anti-war sentiment and disillusionment in the late 1920s and 1930s.

Haig described Lloyd George as 'shifty and unreliable'. Peter Simkins described Lloyd George's memoirs as 'the greatest work of fiction in the English language'.[6] Liddell Hart's criticisms of the British high command were based on absolutely no first-hand knowledge: he left the army as a captain and never served in a formation HQ. Overall, their work paints a picture of Haig and his immediate subordinates as butchers and bunglers. Yet it was not so. Haig, his subordinates, and the soldiers they commanded deserve far better.

We can make some high-level observations about the operational level during the Great War. As previously stated, it was not really identified as such at the time. Yet it is clear that commanders and staffs could conceive of, plan and execute both theatre-wide campaigns and major operations with strategic objectives.

They were not always successful. The Western Front was exceptional, not just in 1914-18 but in the whole of military history. On the Western Front the armies of three empires, subsequently assisted by the richest nation on earth, fought for years in a geographically constrained theatre with no open flanks. As we shall see in Chapter 4, the tactical problem of the rifled bullet dominated almost all other considerations. No army made a breakthrough on the Western Front in four years. But commanders could, and did, think at the theatre and campaign level even within those constraints.

By 1916 Falkenhayn had concluded that major operational-level encirclements (as we would now describe them) would not work on the eastern front. There were exceptions, but even given the geography of Romania (where he commanded an army) the enemy got away. The root cause was that horsed cavalry did not actually demonstrate sufficient mobility.[7] That is seemingly paradoxical, but was certainly corrected 20 years later.

---

6 Professor Peter Simkins, personal communication.

7 Norman Stone, *The Eastern Front 1914-17*. (London: Hodder and Staunton, 1975), 184-5.

We should rethink what 'the operational level' means in terms of the Western Front. The Somme and Passchendaele were campaigns: not battles. The terminology is unhelpful. Other 'battles', such as Verdun, had major theatre-level implications. They should also be seen as campaigns or operations, in the sense of 'the operational level of war'.

More broadly, we should see both the Somme and Jutland as clear British operational successes. One side, or both, may or may not have achieved its tactical objectives. For example, at Jutland the Germans did not destroy the British battlecruiser fleet in isolation. Jellicoe did not decisively defeat the German High Seas Fleet. But, as we have seen, in both cases the British achieved significant operational outcomes.

Several writers have not seen that. That reveals two things. The first is that the distinction between 'operational' and 'tactical' is not yet widely accepted. The second is that the same applies to the distinction between losses as outputs (for example, British ships sunk) and consequences as outcomes (that the High Seas Fleet was bottled up in harbour for the rest of the war).

All this helps us understand the events of the Great War. It in no way detracts from the sacrifice and the casualty rolls. But it does, perhaps, helps us realise that it was not all pointless and futile; although that is what it may have seemed at the time, and has largely been portrayed as such ever since.

4

The Great War: Tactics

The 52 months of the Great War saw what was probably the most concentrated, profound period of tactical change in the history of warfare. Not least, technology was applied to develop weapons which were unimaginable in 1914. Yet they were in everyday use, with appropriate tactical methods, well before November 1918. This chapter focusses largely on the tactics of British, Empire and Dominion forces in order to illustrate those processes and highlight important variations elsewhere.

We start in the air. In August 1914 airships were probably more useful than aircraft. The British Royal Navy (RN) had six airships, rising to 103 by November 1918. Naval airships were mostly used for fleet reconnaissance. They operated both from shore bases and ships at sea. They were increasingly used for anti-submarine reconnaissance and even attack. The RN was well aware of the threat posed by German airships ('Zeppelins'). In 1914 it conducted a small number of bombing raids, by aircraft, against Zeppelin hangers at Dusseldorf, Cologne, and even Friedrichshafen on Lake Constanz.

An aircraft first flew off a ship underway, HMS *Hibernia,* in 1912. The first ship capable of both launching and recovering aircraft was HMS *Furious,* which became operational in 1917. The RN had ordered a torpedo-bomber seaplane in 1913, but it was not successful. The first air-launched torpedo attack was successfully carried out by a RN aircraft in the Dardanelles in August 1915. By the end of the war the RN had ordered 300 purpose-built Sopwith Cuckoo torpedo bombers.

Armies used balloons extensively for battlefield observation, and particularly for the control of artillery fire. Tethered balloons became very vulnerable in 1916, once airmen learnt how to destroy them using rockets and phosphorous bombs. Armies initially used aircraft for reconnaissance and observation. The first British soldiers to sight German forces were Royal Flying Corps (RFC) aircrew. An aircraft sighted German troops crossing the old Waterloo battlefield whilst the BEF deployed at Mons, 25 miles away, on 22 August 1914.

There was no policy to arm aircraft in August 1914. However, the RFC had some Lewis (light) machineguns available. They were soon mounted on aircraft. Specialisation into observation, bomber and fighter aircraft (and units) developed rapidly. It was generally complete by the summer of 1916. Aerial photography and photographic interpretation became important tools for intelligence, mapmaking, artillery fire planning, and counterbattery location.

Aircraft performance improved considerably. In 1903 the Wright brother's aircraft ('Kitty Hawk') had a 12 hp engine. By late 1918 some fighters had 500 hp engines. By comparison, the most powerful tank engines (in the German A7V) could deliver 200 hp. Much of the improvement in aircraft technology was also in place by the summer of 1916.

Control of the air became highly important, principally to allow photographic reconnaissance. It was critical to the outcomes of the battle of Arras and the Nivelle offensive. Bombing airfields, and shooting down enemy aircraft in the air, became important air superiority tactics.

Increasing range enabled aircraft to attack targets well behind the enemy's front line. That became particularly important in delaying the arrival of enemy reinforcements in the early stages of a battle. RFC attacks on troops advancing across the old Somme battlefields was a significant factor in the failure of the German Spring offensive of 1918.

Pre-war RN plans included the provision of antiaircraft guns and fighter aircraft to protect the Thames. Zeppelins raided both Britain and Paris from the early stages of the war. There were 53 Zeppelins raids on Britain: an average of almost exactly one per month. Up to 14 airships took part on each raid. Those raids killed 556 civilians and did £1.5 million worth of damage to property. Industrial production was disrupted, but generally restored within a few days. Counters were developed. The Zeppelins were forced to operate at night, and infrequently, due to losses. 30 of the 84 airships were lost.

The German air force attacked Britain 57 times with long-range Staaken and Gotha bombers from May 1917. The domestic political consequences in Britian were enormous. The raids led to the formation of the Royal Air Force (RAF). 43 German bombers tied up 400 front-line British aircraft to defend London. It was described as one of the greatest diversions of military resources of all times.[1] But, once again (and as predicted by one of the German pilots), heavy losses forced the Germans to operate at night, and then to withdraw the bombers. The RFC, and then the RAF, conducted long-range raids against targets in western Germany. The results were just as poor.

At sea, coal-fired steam turbines had made Dreadnought battleships possible. They were also their main limitation. At speed, battleships used about a ton of coal per mile. They typically carried about 900 tons. Under those conditions their operating range was only about 450 miles. Oil-fired steam turbines were entering service. British oil-fuelled 'fast' Dreadnoughts, such as the *Queen Elizabeth* class, were typically four knots faster. In practice they also had a greater operating range.

The range and seakeeping of destroyers were limited. German destroyers could not reach the main Grand Fleet base at Scapa Flow in the Shetlands. British destroyers could reach German bases near Wilhelmshaven. However, they would have to return to port if the fleet stayed at sea for a second day.

Conversely, cruisers were primarily designed for range and seaworthiness. Their armament varied considerably. Some were armoured; some were not. Battlecruisers had the range and seakeeping of a cruiser, the armament of a battleship and the speed of a fast battleship. Where they came up against cruisers, as at the Battle of the Falklands in December 1914, they were lethal. However, they had roughly the same protection as (armoured) cruisers. Where battlecruisers came up against battleships, as at Jutland, the results were catastrophic.

The RN established a radio listening post on Scarborough Head early in the war. The Russians captured German naval code books, similarly early in the war, and passed them to Britain. German naval communication security (COMSEC) procedures were consistently poor (except in the Flanders Squadron). That gave Britain a significant advantage. That is how the Admiralty knew that the High Seas Fleet was putting to sea on 30 May 1916. That triggered the Battle of Jutland.

Admiral Sir David Beatty was commanding the British battlecruiser fleet of six battlecruisers and four fast battleships. He was ordered to

---

1 John Slessor, *Airpower and Armies*. (Tuscaloosa, Alabama: The University of Alabama Press, 2009), 23.

put to sea from Rosyth, near Edinburgh, by telegraph. Admiral Sir John Jellicoe, commanding the Grand Fleet at Scapa Flow, was also ordered to sea. Jellicoe had 24 Dreadnought battleships and three battlecruisers. The British had a total of eight armoured and 26 light cruisers, and 79 destroyers. The Germans had 16 Dreadnoughts, six pre-Dreadnought battleships, five battlecruisers, 11 light cruisers and 61 destroyers. The Germans did not know that Jellicoe had sailed. Beatty did.

At 1525hrs on 31 August the British battlecruisers sighted the German battlecruiser fleet commanded by Admiral Franz Ritter von Hipper at a range of about 11 miles. The Grand Fleet was 60 miles to the north. Beatty intended to draw the Germans northwards. Hipper intended to engage Beatty until the rest of the High Seas Fleet under Admiral Reinhard Scheer arrived. The battlecruiser HMS *Indefatigable* sank at 1602 hrs. HMS *Queen Mary* sank at 1626 hrs. Both were hit by German battlecruisers and blew up. Hipper lost no ships.

It was not critical. At 1648 hrs the British cruiser HMS *Southampton* accurately reported the location and heading of the German fleet. Beatty and his battlecruisers slipped into place in the van (that is, as the vanguard) of the Grand Fleet at 1845 hrs. HMS *Invincible* had sunk at 1835 hrs. The Grand Fleet was now, however, in a perfect position. Jellicoe's Dreadnoughts were in line astern. Scheer was sailing in a single column straight towards the middle of Jellicoe's line. Jellicoe's battleships could all engage, with their broadsides. Only Scheer's leading ships could engage; and only with their forward turrets. Jellicoe had 'crossed Scheer's 'T''. It was a Nelsonian dream come true.

However, Jellicoe did *not* order his fleet to close with, and destroy, the enemy. The battle was tactically inconclusive. Writers have made much of the fact that Jellicoe could have lost the war in an afternoon. The truth is simple. At Tsushima, in 1905, the Russian fleet had lost several ships to Japanese torpedoes. Torpedoes had an effective range of about 4,000 yards. That was about the same as that of the Russians' gunnery rangefinders. It made some sense to turn battleships *away* from destroyers in order to keep outside torpedo range. However, by 1916 gunnery ranges were much greater. Torpedo ranges were not. Jellicoe turned the Grand Fleet away not once, but twice. He was following the RN's Fighting Instructions. Scheer got his fleet away, substantially intact.

Destroyers carried torpedoes with a 4,000-yard range. Dreadnoughts had secondary batteries (typically of four- or six-inch calibre) which could wreck a destroyer at 10,000 yards. That was, basically, what secondary batteries were for. It was, of course, not that simple. Not least, a ship crossing at a constant speed was a predictable target for a torpedo. A ship head-on was a far smaller target, and one which could dodge reasonably well. After Jutland, Fighting Instructions were rewritten to instruct commanders to turn *toward* a torpedo threat. At Jutland, the doctrine (Fighting Instructions) was behind the technology.

Most of the naval war was not about fleet actions: it was about the attack on trade. U-boats had a reasonable operating range but were slow. They could make perhaps eight knots on the surface and five submerged. They could not remain submerged for long. If submarines could cruise on the surface and pick off merchantmen with gunfire and little fear of reprisal, they could be very successful. Convoying made targets much harder to find, but provided better pickings if the convoys were found. They were, however, also escorted; and therefore better protected. Airships or tethered balloons made daylight attack by U-boats risky. A four-inch gun on an escort could sink a submarine with one hit. Convoys sailing with few lights at night would be very hard to spot. In practice, the introduction of convoying defeated the U-boat. They continued to put to sea, in ever greater numbers, but to ever less effect. After convoying was introduced merchant shipping losses dropped to under one per cent of sailings.

The loss of the three British cruisers to one U-boat on 24 September 1914 alerted the British public to the submarine threat. Three British pre-Dreadnought battleships were lost to torpedoes off the Dardanelles in two weeks in May 1915. Those six ships represent almost half of the 14 battleships and cruisers lost by the RN to submarines in the whole of the war. The submarine was not yet the threat it would become, but the lesson was learnt.

On land, the German Army crossed the Belgian frontier on 3 August 1914. It came under fire from the forts around Liege the next day. German 15cm howitzers and 21cm mortars did some damage to the forts. On 11 August German super-heavy 28cm and 42cm artillery arrived.

However, the Liege forts had been designed in the 1880s. Their concrete protection was unreinforced: reinforcement came into use just a few years later. The concrete was badly poured, resulting in structural weakness.

The forts were poorly ventilated, and their latrines were inadequate. They became uninhabitable after a few days' siege.

'Ring' fortresses, consisting of several small forts surrounding a city, were designed with relatively little protection on the rear face of each fort. That was intended to prevent a fort which had been captured being held against counterattack by the garrison. Liege was no exception. However, the outer defences had gaps which allowed the Germans to cross the River Meuse above and below the city. The inner defences were poorly maintained: attacking infantry could, and did, penetrate between the forts. The Germans then brought their heaviest artillery into the city and destroyed two of the forts *from the rear*. The rest, badly damaged, surrendered.

In 1914 Przemysl in Poland was just inside Austria-Hungary, on the border with Russia. It was a fairly modern ring fortress. It was besieged by the Russian Army on 16 September 1914. Initial attempts to storm Przemysl cost the Russians 40,000 casualties in three days. The siege was briefly lifted, and then re-imposed through the winter. In March 1915 Russian heavy artillery enabled the northern forts to be stormed. The garrison then destroyed everything of military value and surrendered. The siege had lasted six months. 117,000 Austro-Hungarian troops went into captivity.

Some of the French forts at Verdun had been modernised. That included adding an extra layer of concrete. They were effectively invulnerable to German siege artillery. Most of the forts' heavy guns had been removed. In February 1916 Fort Douaumont had a garrison of just 56, against an establishment of 500. Unsurprisingly, a determined infantry assault eventually took Douaumont. Fort Vaux eventually fell with the loss of about 20 French dead and 100 wounded. The attackers lost 2,742 casualties against Vaux alone. The French recaptured Vaux with the help of a 400mm railway gun firing against the rear face of the fort. Engineers then brought the fort back into commission whilst the battle continued outside. They dug new tunnels 17 metres below the existing galleries. The new works strongly influenced the design of the Maginot line in the 1920s and 30s.

In late 1914 German engineers began underground mining against front-line trenches on the Western Front. Ten mines were blown against Indian troops around Ypres on 20 December 1914. Tunnelling and mining became major features of trench warfare. It gradually moved away from offensive mining to underground construction. For example, an extensive system of deep shelters was built in the Ypres salient after the Passchendaele campaign. Britain raised 28 tunnelling companies, each of 274 officers and men. Infantrymen were also allocated for labouring duties. The BEF deployed up to 15,000 men working on tunnelling projects at any one time.

The German drill regulations of 1906 required a regiment of about 3,200 men to attack in a formation 700m wide and 1,000m deep. The front rank would be spaced one to two meters apart. The leading companies therefore had an effective density of about one man per meter of frontage. A regiment had an effective density of four to five men per meter. Seen by the enemy at ground level, they would look like solid columns. In 1914 the BEF simply shot them flat. The Germans had made some allowance for modern small arms fire, but overlooked:

- the 'searching' effect of high-velocity long-range, flat-trajectory rifle fire. In practice, bullets will hit men standing upright over a wide range bracket.
- the British practice of using 'combined sights', right down to section level. They deliberately distributed their fire throughout the depth of an attacking column.
- rapid fire. The 'Mad Minute' practice in the British Musketry Regulations of 1909 required 15 rounds to be fired in one minute. That was routinely exceeded. The record was 36 *hits* at 300 yards.

The Germans were relatively inexperienced in the conditions of modern warfare; British units less so (largely due to the Boer War). The doctrine of several European armies (including the German) relied greatly on rationalistic logic and aspiration. The British Army relied more heavily on empirical evidence. Researchers generally do not seem to have looked at the German regulations and calculated the effective ratio of force to space.

Trench systems were rudimentary at first. The trenches were not joined up across the Western Front until the spring of 1915. The winter of 1914-5 must have been miserable for the soldiers involved. Front-line fire trenches were gradually thickened up with support and reserve trenches. Communication trenches linked them from front to rear. Designs became extremely sophisticated. Traverses in the forward trenches prevented the blast from shell fire travelling along the trench. Communication trenches were zigzagged to protect against long-range machinegun fire.

Troops in trenches are vulnerable to high explosive from above. That might be from hand grenades, rifle grenades, mortar bombs or howitzer shells. All four were quickly developed or procured. British units did not have a good supply of reliable hand grenades until well into 1915. That put the soldiers of the Australian and New Zealand Army Corps (ANZACs) at a major disadvantage at Gallipoli, for example. The German Army had both grenades and trench mortars, which they initially saw as siege weapons. The biggest trench mortars had very short ranges: perhaps a few hundred yards. However, their bombs were massive. They could completely wreck a section of trench. The mortars were also relatively immobile and vulnerable to counterbattery fire.

The early battles of 1914 showed that it was suicidal for field artillery to fire over open sights, due to enemy small arms fire. The guns were soon withdrawn behind cover, typically 2-3,000 yards back, and linked to forward observers by telephone. Barbed wire was quickly put up in front of trenches to stop attacking infantry. Dugouts were built into the front faces of trenches to provide shelter from both artillery fire and the weather. Reinforced concrete was used in some front-line dugouts by the spring of 1915. Machineguns were sited well forward, and fired in enfilade across the front, to form a curtain of defensive fire.

The resulting combination was lethal. The barbed wire held up the attackers, who were then cut down by machine guns. Defending artillery decimated any attackers caught in the open. The attacker's field artillery could not easily cut barbed wire. The defender's field batteries were relatively safe on reverse slopes to the rear. The key problem, however, was the effectiveness of the fire of unsuppressed (or un-neutralised) rifles and, particularly, machineguns.

At Neuve Chapelle in March 1915 the BEF found that it was not difficult to storm a single trench line. A short, heavy bombardment neutralised most of the defenders' machineguns. A quick assault would then clear the position, albeit at some cost. However, surviving machineguns (often on the flanks) made it impossible for the attackers, or subsequent waves, to penetrate to depth. Breakthroughs became virtually impossible.

German defences developed both in depth and complexity. By early 1916 they consisted of three separate positions, each of three trenches. The total depth was about five kilometres. The second position was explicitly intended to protect the field artillery. It also stopped any attackers who breached the first position and provided shelter for reserves. Barbed wire was thickened into several belts each 40 yards or more deep.

Attackers rapidly developed tactics to work along trench lines. They used hand grenades and bayonets to clear traverses one by one. The same tactics were also used by counterattacking troops. This 'bombing along traverses' became time-consuming and inconclusive.

Medium artillery, typically of four to six inch calibre (105 to 155mm or so), was increasingly used for counterbattery fire to neutralise defending field batteries (of 75mm, 77mm or 18pdr (84mm) calibre). Specialist counterbattery staffs were introduced, typically at corps HQs. Counterbattery fire required a variety of locating techniques, as well as accurate mapping and aerial photographs. That in turn drove the need for control of the air. By 1918 aerial observers could direct counterbattery fire in near-real time. Antiaircraft artillery developed slowly.

Gas was first used by the Germans in the Second Battle of Ypres in 1915. Initially it was released from canisters in the attacker's front line. Subsequently it was delivered by artillery shells. Smoke shells were also developed and brought into use.

The BEF grew rapidly through 1915 and the spring of 1916, when it reached a strength of about 60 divisions. Its attacks in 1915 (Neuve Chapelle, Aubers, Festubert and Loos) were all planned to form part of coordinated Allied offenses. They can all be seen as failed attempts to break through increasingly sophisticated German defences. They can also all be seen as part of a process of a new army learning by fighting, and learning in the utterly novel conditions of the modern battlefield. Cavalry was held ready to exploit any breakthrough, but none was achieved. In defence, cavalry formed a screen behind infantry positions as a form of counter-penetration. They were rarely needed.

Wider attack frontages were needed. That would prevent enfilading machineguns (whose fire might reach up to 3,000 yards in from each flank) interrupting the main attack. General Sir Henry Rawlinson had recognised that in 1915. Wider sectors meant bigger attacks and more guns. The French Army had entered the war with very few modern medium and heavy guns (which was why many French forts were disarmed). Their older guns lacked proper recoil mechanisms. They had to be hauled back into position after each round, so their rate of fire was slow. British planners mistook total *weight* of fire for *rate* of fire (intensity). For both reasons, preparatory bombardments got longer and longer.

Armies developed their tactics and revised their doctrine. In the early summer of 1916 French tactics were better than British. Neither army had worked out how to break through a thoroughly well-prepared defensive position. That was also true at Gallipoli and on the Isonzo. The Germans achieved some success in the east against the Russians, the Serbs and the Romanians, all of whose defences were less well developed.

Haig was the operational commander on the Somme. It was entirely right of him to seek a breakthrough, if it was possible. Why would he not? It was not, however, one of his operational objectives. It was also right for Rawlinson, the relevant Army commander, to seek achievable *tactical* goals. A so-called argument between Haig and Rawlinson concerned the difference between a depth of 2,500 yards (Haig) or 1250 yards (Rawlinson) as the objective for the first day. They compromised. Neither actually planned to break through the whole position that day. The German defences consisted of three separate positions (although the third was not yet complete). South of the Albert-Bapaume road the objective for 'Z Day', 1 July 1916, included the first position and up to the beginning of the second. North of the road, the objective included all of the second position. Orders were written, however, to be able to exploit a breakthrough if it did occur.

Z-Day was not entirely a ghastly failure. It was far more successful than the first day of the previous British attack, at Loos. South of the road, the French Army and two British corps seized most of their objectives, at reasonable cost. The disaster occurred north of the road.

In some areas in the north the bombardment was planned to lift 10 minutes before the infantry attacked (to allow the debris from mine explosions to settle). In those sectors German trenches, and machineguns, were manned before the attackers left their trenches. Conversely, in some divisions the attacking troops were in no man's land before the artillery lifted. Some got to the enemy front line before the defenders emerged. In one division, however, no officers above company level went forward in the assault. Thus even quite promising successes were quickly defeated by German counterattacks. In some places the attackers were held up by machineguns to their flanks, because the attack there had broken down. Not enough counterbattery work had been done, so German artillery caused casualties in the supporting waves even behind the British front line. Overall, it was the greatest loss of dead and wounded in one day that the British Army has ever suffered. The total was about 55,000 men killed, wounded and missing in action.

However, the British continued to attack. 46 minor operations were conducted to prepare for the next major attack. On 14 July the BEF seized 6,000 yards of the second position, just south of the main road, for modest losses. A warning order was released on 18 July for the next major operation. There were then four separate battles, consisting of 90 local attacks, before the battle of Flers-Courcelette on 15 September. That day 4,500 yards of the third position was seized. Tanks were used, successfully, for the first time. There were then six further battles up to 18 November. Haig reported successes, such as the seizure of Beaumont Hamel, at the Allied conference at Chantilly on 15 and 16 November. Those battles also captured the whole ridge and allowed the preparation of good defensive positions for the winter. See Figure 4-1.

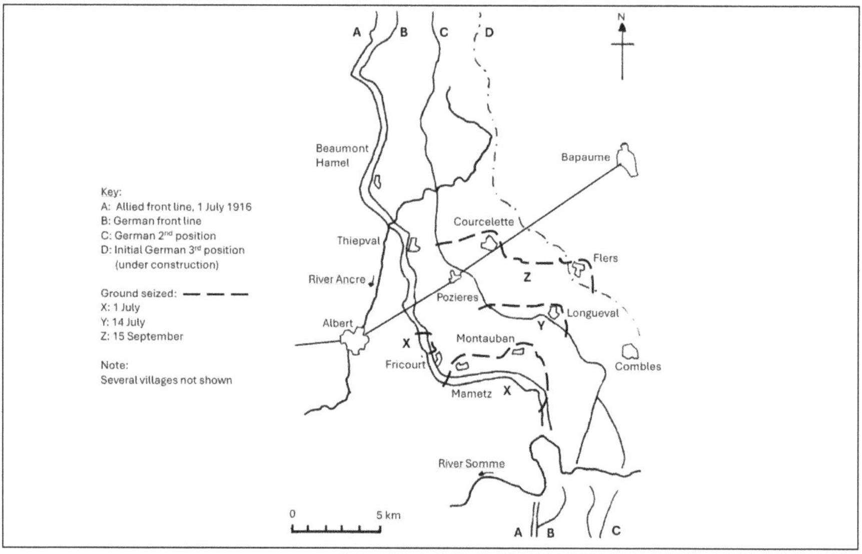

Figure 4-1: Ground Seized by the British Army on Selected Days on the Somme, 1916

The initial British objectives lay astride a ridge over 150 metres above sea level. It runs from the northwest, above Thiepval, southeast almost to Combles. From the German second position, around High Wood (just northeast of Longueval) and Pozières, one can see for many miles northeast into the German defences and rear areas. Haig, Rawlinson and their staff would have known that from map study before the campaign. Ludendorff issued new orders on 15 September, the first day of the battle of Flers-Courcelette. Those orders were to prepare what would become the *Siegfried Stellung* (wrongly known to the Allies as the Hindenburg Line). The withdrawal would be known as Operation *Alberich*. The British campaign on the Somme had a major operational impact.

British tactics had developed enormously. Rawlinson had wanted to form up the attacking infantry in no-man's land before dawn on Z Day (1 July). The French vetoed that as being too complicated. It worked spectacularly well on 14 July. Night attacks became common, and often succeeded. The preliminary bombardment on 14 July was, effectively, 16 or 17 times more concentrated than it had been for Z Day. An artillery officer, Major Alan Brooke (later Field Marshal Viscount Alanbrooke), had studied French artillery methods at Verdun. He then developed 'barrage' techniques. A barrage was, literally, a deep curtain of fire which moved forward in front of the attackers. Counterbattery techniques were improved. Machinegun barrages were fired over the heads of advancing troops to prevent early counterattacks. The infantry's advance was coordinated with the barrage and the tanks. Tactical objectives were consolidated using 'bite and hold' tactics. They enabled a position to be seized, consolidated, and held against counterattack. The process was repeated a few days later. 'Bite and hold' had been identified by Rawlinson after Neuve Chapelle in March 1915.

Many writers have written at length and with conviction about the Somme without ever having visited the ground, nor even studied maps in detail.[2] The same applies to many other battles. The study of warfare should, surely, be more than just the comparative study of written sources. Similarly, many writers don't really consider what the protagonists were actually trying to achieve, and measure progress and outcome in those terms.

The British Army did not have enough guns until the end of 1916. An attack needed roughly one field gun per 15 yards of front. It needed one medium battery to neutralise every defending field or medium battery. It needed enough heavy guns for destruction and interdiction targets. That

2 I have visited the Somme six times. I have facsimiles of contemporary large-scale 'trench' maps. I also have modern large-scale maps.

meant several dozens, or hundreds, of medium and heavy guns. The key was to assemble them, and register them, without being detected. At Arras in April 1917 (nine months after the beginning of the Somme campaign), the British concentrated 1,400 *more* guns. 500 were medium or heavy. At both Cambrai (in late 1917) and Amiens (in August 1918) the build-up of guns was undetected before Zero Hour.

Ludendorff and Hindenburg oversaw a revision of German defence tactics. It started during the later stages of the Somme offensive. The front line was held far more sparsely. It did not rely on the defence of a continuous trench line. The main defence was to be waged as a series of counterattacks in the battle zone, which was *behind* the front line. The front line became an outpost line. Thus German tactics were organised around combinations of outposts, a battle area, and reserves.

The Germans had already formalised a process of stationing reserve divisions behind a threatened front. Army Groups were formed. They allowed the roulement (rotation) of divisions to be conducted more effectively. The Somme front was held by eight German divisions on Z Day, but by the end of the battle 89 further divisions had been employed. At the end of August the Germans had only one division available in reserve, anywhere. The six divisions that were sent to other fronts were all exhausted after fighting on the Somme. Four other divisions were broken up.

The Germans formalised the use of reverse slopes in defence. In Flanders, in particular, the reverse slopes were extremely subtle. An elevation of perhaps five metres was enough to conceal a defensive position. Pilckem Ridge, the objective for the first attack of the Passchendaele campaign, is less than 15 metres high.

Casualties on the Somme were heavy. The ANZACs lost as many men in six weeks on the Somme as they had in eight months at Gallipoli. The German Army lost more men on the Somme than at Verdun: perhaps 100,000 more. When they assumed command in September 1916, Hindenburg and Ludendorff were not convinced that the German Army was capable of defending successfully in the West. In essence, the Germans could not stop British 'bite and hold' attacks. They could, however, counterattack to regain lost ground.

The circumstances of the Arras and Ypres operations in 1917 were described in Chapter Three. The Ypres campaign did not achieve its objectives. However, the German Army did not conduct any large offensive, anywhere, in 1917. Thus, in simple terms, all the reserves of the entire German Army could be used to stop the British at Ypres. The Germans were unlikely to run out of troops. They were, however, pushed back. The British were within a few hundred yards of their objectives when the operation was closed down in November. They were so close that a further, limited operation by four brigades was attempted on the night of 1-2 December. Surprise was lost. The attack failed. The British suffered about 1600 casualties. The attempt was not repeated. The Germans and Russians were already negotiating at Brest Litovsk.

On 20 November six British divisions, supported by 476 tanks, had attacked near Cambrai. There was no preliminary bombardment. A massive fireplan started the moment that the assaulting troops crossed the start line. The tanks were already moving forward. Surprise was complete. Attacking infantry used Lewis guns to suppress isolated pockets of defenders, and mortars to neutralise surviving machineguns. By 26 November the British had advanced about four miles on a ten-mile front.

Operational surprise was complete. No German reserve divisions were available for immediate counterattacks. A deliberate counterattack started on 30 November. It regained about two miles by 3 December. The British withdrew to a defensive line, which included much of what had been the German third position, on the night of 4-5 December. The battle was closed down.

Why was Cambrai fought? There are very few clues. It may have been an experiment to try out the use of tanks *en masse*. The experiment cost about 44,000 casualties and gained little ground of any significance. Reflection suggests that it was indeed a trial, but on a different scale. Did the BEF now know how to attack through German defensive positions in depth? The answer was emphatically 'yes'. Could the BEF achieve complete operational and tactical surprise? Yes. Cambrai would be the last chance to confirm that before the German Spring offensive in 1918.

The Western Front represents almost four years of stalemate, but not stasis. Much of the tactical problem was that the Germans also adapted and developed their tactics. By the end of 1917, the German Army was not just far better at defending than it had been in 1914, it was also far better at attacking. They called their new assault procedures 'stormtroop tactics'. Ironically one German POW referred to them as 'English tactics'. The generic term is 'infiltration'.

Infiltration tactics were probably the most important tactical development of the Great War.

The British Army rarely used the term. However they adopted infiltration methods up to at least battalion level for decades to come. The German Army excelled at them in the Second World War. We will return to the topic of infiltration several times in this book.

German tactical successes in the Spring of 1918 were due to several factors. Stormtroop tactics were one. Short, intensive, surprise bombardments with a very high proportion of gas and smoke shells, was another. British political decisions also contributed.

In October 1917 the British Cabinet ordered Haig to take over more of the front line from the French. In December he was ordered to reduce divisions from 12 battalions to nine to economise on manpower. Haig also had to send divisions to Italy, although some returned. So, slightly fewer divisions had to hold appreciably more front line, with only three-quarters as much infantry per division. Divisional defensive schemes were now impractical: there were not enough battalions to hold reserves for counterattacks at all levels. Frontages were excessive: one battalion is reported as having to hold 6,000 yards. 1,000 yards would have been typical a year earlier. The Germans' tactical success on 21 March 1918 was almost inevitable, and much of the reason lay in political direction.

10 British divisions suffered particularly heavily over the next few weeks. Some continued to fight, and were not withdrawn until May. Tens of thousands of reinforcements, previously withheld by the War Cabinet, were sent out from England between March and June.

On 8 August 1918 the British Fourth Army and part of the First French Army attacked at Amiens. The tactics were largely those of Cambrai. Within hours the Australian and Canadian Corps, in the centre, had broken through the whole depth of the German position. A German corps HQ was overrun, probably for the first and only time in the war. (The commander of the tank which seized it had been explicitly ordered to that location. Its position seems to have been detected by radio intercept, or similar.) British cavalry tried to exploit.

However, procedures for committing cavalry through the front line didn't quite work. Unit and brigade commanders were perhaps too cautious. The RAF did good work at first. When it switched to interdiction it was much less successful. It took considerable losses. To repeat: 8 August was the 'Black Day of the German Army'. The German Army was then beaten on the Western Front and the War was brought to an end. Germany lost.

By that stage, British attack tactics were far more flexible. The leading waves of infantry would follow the barrage closely, right up to the objective. That was typically a line just within the limit of field artillery range. By doing so, the attackers would generally break right through the Germans' first and second positions. They would bypass any pockets of resistance, which would then be cleared from the flank or rear by follow-on forces. They did not waste time bombing along the traverses. Once the second position had been taken, the field artillery quickly moved forward. Some reserves would be well forward to defeat immediate counterattacks. The area seized would be thoroughly 'mopped up' by other follow-on forces.

In the South African War Kitchener had forbidden frontal attacks. By 1918 flank or rear attacks were absolutely routine, from section to company level. They were an integral part of attacks up to corps level. When well-planned and executed, such tactics were highly effective. At their best, they could break right through the toughest defences. After Amiens, they generally did.

The Germans almost always counterattacked. By mid-1917 the British could generally beat off immediate counterattacks. Nonetheless, those counterattacks had important second-order effects. Anticipating them constrained the British to limited, bite-and-hold attacks. It also severely limited opportunities for exploitation.

For the attack on the Hindenburg Line on 29 September 1918, the US 27th and 30th Divisions attacked as part of the Australian Corps. Rawlinson widened the attack sector by committing the British 1st, 12th and 46th Divisions on the flanks. He was applying the lessons of Loos, three years before. Ironically it was only the British 46th Division, on the right, which broke through the Hindenburg Line that day. (The US 30th Division exploited the 46th Division's success to capture Bellicourt and Nauroy by the following morning.) The 46th Division suffered just 800 casualties. The 32nd Division then passed through it before midnight. It was one of the greatest feats of arms of the whole war. By late 1918 the British could break *in* to a defensive position quite easily. They could break *through*. But, with one important exception, they never broke *out*.

The British had also attacked on 19 September. Three cavalry divisions were committed after an initial breakthrough by two infantry corps. By about

11am 88 squadrons of cavalry, with supporting horse artillery, were on the move. Within 36 hours the enemy were in full retreat. The cavalry overran an Army HQ 45 miles to the rear. The RAF had done an outstanding job of destroying communications infrastructure. They continued to interdict retreating enemy columns. Within a week, an enemy army group had been defeated. Within a month the enemy had signed an armistice.

'Armageddon' denotes a catastrophe of Biblical proportions. It was a real place in Palestine. Its modern name is Megiddo. The Battle of Megiddo, from 19 September, began the destruction of the Turkish Yilderim Army Group, and hence the Ottoman Empire. Yes: cavalry *could* have operational effect. In Palestine, it did. Writers often denigrate Megiddo, citing this or that reason why it should not be considered to be important. The British Army had thought about its tactics, adapted them to the conditions, and brought about operational and strategic effect. What more should it do?

Why is Megiddo not better known? The answer is probably quite simple. Not a single Regular cavalry regiment fought there. The cavalry was either Indian, British Yeomanry, or Australian and New Zealand Light Horse. First-hand written sources from the Great War contain very little written by Regular cavalrymen in Palestine. There was also very little British Regular infantry. Much of the British infantry had been sent to France and replaced by Indian battalions which had been doubled. There was therefore little or no institutional, Regular memory. The Arab Revolt, which took place more or less simultaneously, might also have been largely forgotten if T E Lawrence had not written his memoirs a few years later.

Haig was a cavalry officer. As chief of staff to (the later) Sir John French he had witnessed, and may have taken part in, the successful British cavalry charge at Elandslaagte in the South African War. Military men are not stupid about things that they have seen. In Libya in January 1916, 180 troopers of the Dorset Yeomanry charged and routed about 1,600 Senussi tribesmen, supported by three Turkish machineguns. The Yeomanry killed between three and five hundred Senussi. Cavalry charges could, and did, succeed. French and Haig were not reactionaries. Haig was a moderate reformer. He knew exactly how difficult it was to create the right conditions for success. As Commander in Chief, he was responsible for achieving an *operational* breakthrough if possible. In the Great War that was only possible using cavalry formations. Haig never threw his cavalry away needlessly. He should not be criticised for trying to use it to best effect.

To summarize: the war saw dramatic changes in the tactics and equipment employed, the techniques of command, and the methods of training. British commanders had no pre-war experience of the operational level of war. They had very little experience *of any sort* above divisional level. They learnt a lot, and they learnt fast. European commanders had far more prior knowledge, and in some cases experience. In some cases they were hamstrung by doctrines which resulted in many unnecessary deaths.

Armies learnt by fighting. So did navies and air forces. War is adversarial and evolutionary. The learning process, in such a large-scale and hard-fought war, was very steep. It wasn't perfect. It was haphazard in some places and also, perhaps, when seen from a distance. Describing it as a learning 'curve' is too prescriptive. But armies did learn. To that extent war is evolutionary, as seen in the lessons which competent armies learn.

Tactically, the German Army could, and did, defeat almost everybody it came up against, early in the war. It repeatedly defeated Russians, Serbs, Romanians, Italians; and the French and the British at the beginning of the war. However, the western Allies learnt. By 1915 the French could defeat the Germans in some circumstances. The German attack on the British at Ypres in 1915 did not succeed. The Germans broadly succeeded in defence, in so far as they contained any breakthroughs in the west, until the end of the war.

Casualties were horrific. If you assemble large armies and have them fight at close quarters for years, they will be. If your armies aren't very proficient, like the Austro-Hungarians, Italians or Russians, they will be. If your army is too small and its replacements are half-trained at first, like the British and Americans, losses will be horrific. If your operational approach is initially one of all-out attack at all costs, like the French, the cost will be high. If your tactical doctrine is like that, your losses will be high until you learn better. Thus the German Army's experience.

The apparent futility of the Great War emerged to some extent from war poetry. It was the work of sensitive young men exposed to horrors which no-one should ever have to live through. 72 published British war poets died in the War. Seven died on the first day of the Somme alone. The poetry of the Great War is a significant addition to the canon of English literature, perhaps even on a par with Shakespeare. But it is *not* history.

The notoriety of the Schlieffen plan largely resulted from German domestic propaganda. The battle of the memoirs was perhaps even more important. Regrettably, Lloyd George won his. Churchill did not forget that. The 'butchers and bunglers' theory is due in large part to Lloyd George, aided and abetted by Liddell Hart. The narrative becomes hugely important. That was never more true than in the history of the Great War.

Taking the Somme as just one example, the relationship between casualties and outcome is not straightforward. Whatever we think about the Somme at the tactical and operational levels, we need to think clearly. Did the British *not* win because of the high casualties? Did they win *despite* those casualties? Greater clarity is required. As described in the Introduction, we need to separate outputs, especially casualties, from outcomes. And we need to see both in terms of the objectives; that is, the goals sought.

The tactical lessons of the Great War have little direct relevance in the twenty-first century. However, they formed a basis for developments later in the twentieth century. They also tell us a lot about innovation, adaptation and evolution.

5

1918-1939

There were several significant conflicts between the First and Second World Wars. The period should, however, be seen primarily as one of transition, innovation and change. In turn military innovation should be considered to be a consequence of the interaction between the strategic situation, technology, and the wider human environment.[1] That environment must include political-military relations. Here that inevitably includes the issue of government funding.

The interwar period is often noted as one of significant military thinkers, across several aspects of warfare. They include Soviet operational theory, so-called 'strategic' bombing and armoured warfare. We shall see that other aspects, not necessarily the province of famous thinkers, also saw significant development during this period. That is probably a consequence of the vast scope and scale of the Great War.

Chapter 5 steps through the strategic situation, technological development and the wider human environment. It then considers warfare on the eve of the Second World War and makes observations and deductions.

We start with the strategic situation. Looking back almost 80 years since the end of the Second World War, we should remember that there were just 21 years between the two World Wars. We can usefully consider three overlapping phases: disarmament; austerity (together with disillusion and depression); and rearmament.

The Paris Peace Conference of 1919 led to five treaties. One was imposed on each of the 'belligerent' nations (Germany, Austria, Hungary, Bulgaria and Turkey). Germany, and particularly its foreign office, then set about deliberately undoing the consequences of the Versailles Treaty. It had considerable success, largely unnoticed by western historians. It was

1 Williamson Murray and Allan R Millett eds, *Military Innovation in the Interwar Period.* (Cambridge: Cambridge University Press, 1996), 1-5.

ably assisted by the British radical economist John Maynard Keynes. His book 'The Economic Consequences of the Peace'[2] argued that punitive reparations imposed on Germany would be counterproductive. It was published within months of Versailles. However, contrary to German claims, the Versailles Treaty did *not* ascribe guilt for the war to Germany. It ascribed *responsibility*. Furthermore, in practice Germany paid little in reparations. It did suffer hyperinflation soon after the war. That was largely a result of German economic mismanagement. However, in popular opinion it was easily taken as proof of Keynes' ideas.

The Paris Conference also established the League of Nations. The League was a very well-meaning attempt to preserve peace, and particularly to improve the way in which international disputes were brokered. It failed for two related reasons. The first was that the United States Senate would not ratify the League's Convention. America did not become a member. Secondly, when faced with Axis aggressions in the 1930s, the League's institutions were not robust enough to prevent war.

There is always unfinished business. In Britain's case Irish independence re-emerged as an issue within two months of the end of the Great War. The brief but sometimes vicious Irish War of Independence was followed almost immediately by civil war in the new Republic of Ireland. Independence for India bubbled up and would not be resolved until after the Second World War. France had similar problems with Indochina; America with the Philippines.

Italy and Japan's expansionist aspirations were not resolved by the Great War. In Italy's case popular opinion felt that its sacrifices had not been rewarded. Japan and Italy would turn German aggression in Europe into a second global war. In Japan's case that meant war in East Asia and the Pacific. For Italy it was in the Mediterranean and Africa.

Ironically, the Paris Conference created numerous new problems in Europe, and some elsewhere. Most of the new states in central and eastern Europe inherited potentially fractious ethnic minorities. There were dozens of border disputes. Several were resolved violently. For example, Poland fought four wars before 1925. Several of those newer issues (such as Romania's claims in Bukovina and Transylvania) were not resolved by 1939.

---

2 John Maynard Keynes, *The Economic Consequences of the Peace*. (London: Macmillan, 1919).

The Spanish Civil War was unusual among the conflicts of the period. Its origins were largely domestic and unrelated to the Great War. However, it took place within the period of international rearmament after 1933. It pitted a right-wing faction against a centre-left government. The former attracted support from Fascist Germany and Italy; the latter from liberals and the Communist Soviet Union. It was a big war: both sides raised armies of over 700,000 men. Germany, Italy and the Soviet Union deployed advisers and some troops. Germany and Russia explicitly used the war as a testbed for their armed forces.

A major failing of the Paris Conference was that it did not address domestic issues inside the five belligerent nations. Why should it have? The key consequence was that Germany was left with domestic political institutions which could not, and did not, prevent a minority political party seizing control in 1933.

A German chancellor, Adolf Hitler, started the Second World War (he was actually born in Austria). Germany was not simply 'unlucky' to experience Hitler's rise to power. There are violent extremists in every society. In 1923 Hitler was jailed for an attempted coup (the 'Beer Hall Putsch') on a charge of high treason. Yet in practice he would become Chancellor in 1932 and assume dictatorial powers in 1933. The immediate circumstances were those of the Great Depression (discussed below). The rise of Hitler and the Second World War did not, however, *result from* the Great War. It was started by a violent extremist who had exploited weak legal and constitutional institutions. The Great War was, nevertheless, an important part of the context.

The six years from 1933 were a period of growing tension, attempts at appeasement, and rearmament. Britain for example, started building warships again in 1936. Regretfully for the most part, the World went to war once again in 1939.

We now turn to technology. Aircraft made of wood and fabric aren't very durable. Nor can they handle stresses resulting from high speed and vigorous manoeuvring. So the interwar years saw a search for new lightweight metals and alloys, together with ways to build aircraft with them. The result was 'monocoque' structures, in which the framework and the skin are a single, integrated whole. The new materials allowed aircraft to become bigger, faster, and carry greater loads. Multiple guns in the wings, and powered turrets, were in (or entering) service by 1939. The epitome of the new aircraft types was the single-engine fighter, such as the Messerschmitt Me 109 and the Spitfire.

By 1939 almost all warships were oil-fired. Protection (in terms of armour, layout and fire suppression systems) had improved. So had the elevation of the main armament and fire control systems. That led to increased range and greater accuracy. Antiaircraft armament was being increased rapidly, and some specialised antiaircraft cruisers were being built. The detection of submarines was improving with developments to acoustic location. The British called it ASDIC. The term was later replaced by the American SONAR.

Naval aviation was developing rapidly. By 1939 purpose-built aircraft carriers could transport, launch and recover modern, high-performance aircraft. The Japanese pioneered the design of amphibious shipping with enclosed docks for landing craft. The US Marine Corps made two other developments. The first was landing craft with ramps ('Higgins boats'). The second was tracked, amphibian assault landing craft. Japanese and American armed forces also gave considerable thought to the tactics and techniques of amphibious warfare.

On land, there had been significant improvements to tanks. By the beginning of the Second World War the best designs weighed perhaps 15 tons and had power-to-weight ratios of about 15 or 20 to one. They had guns of 35-40mm calibre mounted in three-man, fully-rotating turrets. Armour was about 15-20mm thick and cast or welded, rather than rivetted. However, very few tanks were that good. British tanks had two key weaknesses, which would take several years of war to overcome. One was that their guns could fire solid shot, but not high explosive shells. The other was a lack of high-power, compact tank engines.

In 1918 radio was still in its infancy. Almost all warships were equipped with radio, as were some aircraft. On land radio had not generally penetrated below brigade level, largely due to the size and weight of the sets. However, technical development was rapid. By 1939 it was even possible to broadcast moving images, with accompanying sound, by radio-frequency transmission. It was called television.

By then it was also possible to transmit radio messages around the world, and to mount radios in tanks and aircraft. Britain was the world leader in one key application: radar. Britain had developed a substantial radar network for the air defence of Great Britain. Royal Navy battleships and cruisers were equipped with radar for surveillance and, increasingly, for fire control. The first surface-search radars were being mounted in carrier-borne aircraft. The first use of radar by the Army was to direct antiaircraft guns.

High-velocity antiaircraft cannon were making low-level air attack dangerous, both at sea and on land. Cross-country trucks were appearing in numbers. The British replaced six-horse artillery gun teams with a seventy-horsepower four-wheel-drive truck. It was clearly a great improvement.

So how did the strategic situation and developing technology interact with the wider human environment? Money was a major factor. The Great War had been immensely expensive. Many countries were short of money. Germany had been effectively disarmed, so western governments could assume that there would be no major conflict in Europe for some time. Britain, for example, did not formally abandon its 'Ten Year Rule' (which assumed no major war for that period) until 1932. At that point Britain's defence spending was less than one seventh of what it had been in 1919-20.

The Great Depression, which started in 1929, followed on from disarmament. Nations reacted differently. Britain's GDP had recovered by 1933; Germany's by 1934 or so. America's had not recovered by 1940. Given a fairly benign international security environment through the 1920s and into the 1930s, defence budgets had been reduced significantly. In particular, there was little money for military innovation.

Countries which had navies were well aware of the high cost of warships, and keen to avoid another naval construction race. The Washington and London Naval Treaties of 1922 and 1930 were fairly successful. They generally maintained a naval status quo, whilst avoiding the need to build new battleships, until the mid-1930s.

The naval treaties limited Japan to nine battleships. The United States was limited to 15, which were split between the Pacific and the Atlantic. In both cases aircraft carriers were intended to achieve air superiority, and then attack the enemy's battleships before a decisive fleet engagement. Consequently American carriers were optimised to launch large numbers of aircraft quickly, operating off the flight deck.

Germany didn't have a navy at all until it started to build warships again in the mid-1930s. Thus the operational problem for the Royal Navy would probably not be an enemy line of battle. It was the threat of single capital ships operating against shipping in the North Atlantic. The solution lay in a combination of cruisers (to find the German raiders), aircraft operating from carriers to damage and slow them, and battleships to sink them. Ship design reflected the Royal Navy's experience in the First World War. Its carriers were designed to operate aircraft from hangars below armoured flight decks. Great care was taken to prevent aviation fuel fires. Hence in both the Royal and the US Navies, the design and use of carriers reflected their particular strategic situations, hence their particular operational problems, and the tactics intended to meet them.

In the late 1930s disillusionment and pacifism were common. Millions of men had fought. Millions had died. The gains, where they occurred, were generally not obvious. In many countries much of the population really did not want to fight another war. That sentiment prevailed in several nations right up to 1939. Witness, for example, Neville Chamberlain's 'scrap of paper' from Munich in September 1938. In western nations, once war came to be seen as inevitable it was met with weary but determined resolve.

Lack of money and popular will did not stop people thinking about the next war. Perhaps the most profound thinking occurred in the German Army. In 1919 its head was General Hans von Seeckt. He commissioned far-reaching studies into all aspects of the German Army's experience in the Great War. The Wehrmacht's successes of the early part of the Second World War resulted in very large part from von Seeckt's studies and the reforms that followed them.

On land, the greatest single lesson from the War was the need for surprise. With surprise, attackers would often seize their objectives in major offensives. Without it, they almost never did. It was almost that simple. Furthermore, surprise meant that the defenders' reserve divisions were often in the wrong place. So, towards the end of the Great War, it had become increasingly likely that the attackers would hold the ground they had seized.

The other big lesson which those who had fought on the Western Front would probably agree on was the need for infiltration. In one form or another it dominated military discussion in the 1920s. Liddell Hart's first major contribution to military thought concerned infantry infiltration tactics, which he codified.[3] French, German, British and American armies

3 B. H. Liddell Hart, *A Science of Infantry Tactics Simplified.* (London: W Clowes, 1926).

all generally adopted infiltration-based tactics. The Russians, hence the Soviets, never did. It is arguable whether they yet have.

That use of sections and platoons to find and attack the enemy's flanks and rear was formalised in several armies before 1939. Infiltration might be part of a much larger operation. However, the word 'infiltration' was not necessarily used. It did not appear in the British Field Service Regulations nor 'Infantry Training'.[4] The long-term result was, in the British case at least, that the lesson was largely forgotten by the end of the century.

Consideration of armoured warfare generally came later. Guderian was not the only, and not the most important, German armoured theorist. He is, however, the most well known in the west. The British Colonel J F C Fuller and Liddell Hart wrote extensively about armoured warfare. They did not have much impact on German development, and their impact of British armoured thinking was limited and sometimes flawed. For example, the choice of gun for British tanks reflected Fuller's guidance. By 1939 the German Army was clearly the world leader in armoured warfare at up to, and including, the operational level. That may reflect its Great War experiences in eastern, rather than western, Europe.

Fuller and Liddell Hart extrapolated enormously from the tactics of British tanks on the Western Front. If they had used Megiddo as their inspiration, they might have come to very different conclusions. Seen from that perspective, the Fall of France in 1940 looks remarkably like Megiddo.

Hart, an Oxford graduate, never served above battalion level. He was gassed on the Somme and did not see active service again. He had no insight into the practicalities of command. He seems to have considered that all British Generals were incompetent, which is statistically unlikely. He became famous as a writer. His thinking about strategy (the 'British Way of War' and the 'Indirect Approach') borrowed heavily from other, mostly British, writers. He failed to acknowledge that. His thinking was highly rationalistic, opinionated, and largely failed when exposed to events.

Reflecting on experience in the Russian and Spanish Civil Wars, Soviet tank theorists believed that tanks would have to be fast enough to be able to outrun horsed cavalry, and proof against artillery fire. The result would be the famous T34 tank. The Russian Civil War had seen very large armies operating over very large distances. (It also saw about seven million

4 Tom Wintringham, *New Ways of War*. (Harmondsworth: Penguin, 1940), 28.

casualties). That prompted Soviet thinkers, notably Marshal Mikhail Tukhachevsky and General Vladimir Triandafillov, to think about long-range, theatre-wide land operations. That was the origin of Soviet 'deep operations' theory. By 1941, however, both officers had been purged. So had much of rest of the Soviet military hierarchy. So: the thinking was done, but the Red Army was incapable of turning it into practice; at least at first.

By 1932, the British Army had produced three new editions of the Operations volume of *Field Service Regulations* (and two of the Administration volume), three editions of *Infantry Training* (of which, much of the first was written by Liddell Hart), and two editions of the pamphlet on mechanised formations (the 'Purple Primer'). The Army then commissioned a high-level report on the overall lessons of the Great War, the Kirke Report, in 1932.[5] The most significant tactical issue which it identified was the need for surprise, coupled to the shocking effect of indirect fire. The Kirke Report also stressed the need to break through enemy defences in the attack. As before, that actually meant to break *out*. We shall return to those requirements repeatedly.

The Army then directed Brigadier Archibald (later Field Marshal Earl) Wavell to revise *Field Service Regulations* once more. Wavell was highly perceptive. In 1930 he had written, amongst other things, that:

a. Mechanised forces, in armoured vehicles, were not yet a reality.
b. Motorisation (in the sense of the transport of land forces over long distances) had been little practiced in the Great War.
c. Experimentation into mechanisation had only happened quite recently.
d. There would be technical limits in armoured vehicle design due to the well-recognised conflict between gun and armour. There would be a practical upper weight limit on armoured vehicles due to the road infrastructure.
e. In 1930 Britain had a lead in military capability, and there was every chance of maintaining it.
f. The greater part of the British Army would eventually become mechanised.

5 *Report of the Committee on the Lessons of the Great War* (The Kirke Report). The War Office, October 1932. PRO WO33/1297.

g. 'Strategic' (meaning 'operational') cooperation between aircraft and ground forces had not yet been worked out.
h. That independent attack by air forces was the most difficult and controversial to predict:

    1. There were claims that aircraft had made armies redundant
    2. Aerial bombing had had little effect in the Great War
    3. The accuracy of antiaircraft fire had improved greatly
    4. Aircraft were very vulnerable (once hit)
    5. Separate (that is, independent) air forces favoured technical development.[6]

He was largely correct.

Unthinking, excessively rationalist airmen leapt on to a fairly simple idea. Flying is different, so there should be separate, independent air forces. We can call that 'aerial exceptionalism'. The earliest airpower prophets were the Italian Giulio Douhet and the American Billy Mitchell. Both were court martialled. Douhet had little direct impact on his own service. Mitchell met Churchill and Trenchard whilst returning home to America at the end of the Great War. It was then that he became an ardent exponent of long-range bombing and an independent air force.[7] More generally, this simplistic thinking (aerial exceptionalism) is probably the most wrong-headed and mistaken military thought to emerge in the twentieth century.

As we saw in Chapter 4, the empirical evidence was that, during the Great War, long-range bombing was generally ineffective. Perceptive pilots had foreseen, quite correctly, that counters would quickly be developed. We should also remember that at the time 'strategic' broadly meant 'anything beyond the battlefield'. So much of the discussion was confused. 'Strategic' just meant 'long-ranged'. That could mean attacking (in modern terms) tactical reserves, or operational targets, or targets in the enemy's homeland. One of the more thoughtful books of the period urged the RAF to focus on interdiction in support of the Army. Specifically, the author meant cutting the enemy army off from reinforcement and supply. That author was the

6 Brigadier A P Wavell, 'The Army and the Prophets', *RUSI: Royal United Services Institute for Defence Studies Journal*, 75:5 (1930), reprinted in 155:6 (December 2010), 86-93.
7 Dr Lawrence M Burke II, personal conversation.

then Wing Commander John Slessor. He would become a post-war Chief of the Air Staff. In 1936 he wrote his general conclusion that:

> 'No attitude could be more vain or irritating in its effects than to claim that the next great war - if and when it comes - will be decided in the air, and in the air alone.'[8]

Nonetheless, institutional imperatives prevailed. The interwar RAF, and its American counterpart, preached 'strategic' bombing for a simple reason. An independent air force had to have an independent role. That meant striking the enemy's cities and war production directly. The RAF was, and the US Army Air Corps wished to be, independent. So strategic bombing had to work. Therefore it would work. It helped that many civilians believed that 'the bomber would always get through'. That was despite the (admittedly patchy) evidence that it would not, and the likelihood that counters would be developed.

In Britain, however, political direction generally prevailed. The government preached that German bombers would always get through, unless adequate defences were put in place. It directed the creation of Fighter Command to defend British airspace. Fighter Command became the world's first integrated network of radar, ground-based observers, fighters and command systems. It was almost exactly the opposite of what the RAF wanted. The Air Staff consistently asked for more bombers. Ironically, by 1940 the British government (*and* the RAF!) fervently hoped that they had done enough to ensure that enemy bombers would not always get through.

Germany had looked at its wartime experience very carefully. Von Seeckt's studies had looked at the idea of long-range bombing and largely discounted it. The Luftwaffe was established in 1935 around a core of 400 Great War pilots. Significantly the chief exponent of long-range bombing, General Walther Wever, was not one of them.

So, what was the situation on the eve of the Second World War? From 1933, and certainly by 1936, nations started to prepare for war once again. The German Army had 10 divisions in 1933. In 1939 it invaded Poland with 66. It eventually grew to over 200 divisions. France, Italy and Russia already had large conscript armies. Modernising them and bringing them up to effective standards of training was resisted in part; expensive; and a major

---

8 Slessor, *Airpower and Armies*, 214.

task. Progress *was* made: in 1940 France had four armoured divisions, six light cavalry divisions (which were actually mechanised), and three light mechanised divisions; as well as large numbers of infantry divisions. When Germany invaded Russia in 1941, the Red Army had over 23,000 tanks.

President Roosevelt had a particularly difficult problem in preparing America for war, right up to December 1941. It was not obvious whether the United States *would* go to war. If it did, it was not obvious who it would go to war against. In retrospect, two issues were highly important for the US Army.

The first was to produce enough officers. The peacetime Army had about 14,000. It then incorporated 19,000 from the National Guard and about 180,000 from the Officers Reserve Corps. The latter had trained in university officer training units and were then discharged to the reserve. Eventually pre-war Regulars were outnumbered by 40 to one by other officers. The 180,000 Officer Reserve Corps candidates formed an essential link. Not least, they played a major part in *training* the huge numbers of officers and soldiers that followed them into the Army.

The second critical aspect of American army mobilisation was the Victory Program, initiated in April 1941. It focussed on war industries. Targets were set to support a field army of four million men and to produce 3,400 aircraft *per month*. At its peak the US produced over 8,000 aircraft per month (for the Army Air Force and the Navy).

Britain started to train conscripts and post them to the Territorial Army (TA) in early 1939. The TA was doubled. By 1939 Britain had five Regular and six TA armoured brigades. They were, however, seriously short of tanks, and the tanks were poorly designed. Britain was also critically short of heavy artillery and antiaircraft guns. It would remain short of antiaircraft guns for much of the war. Aircraft carriers and battleships were being built, along with some cruisers. Escorts could be produced faster and were, for the time being, a lower priority. The RAF had good fighter designs and was building them fairly quickly. It had designs for heavy bombers. However it also had large numbers of fairly useless light bombers, and hundreds of medium bombers. It did not train to fly bombers over water, nor at night. Yet it had a well-developed doctrine of 'strategic' bombing. This seemingly obvious discrepancy of aims and means was institutionally ignored

Irregular activity continued. 11 Czechoslovak intelligence officers, headed by Colonel František Moravec of the General Staff, were secretly flown to Great Britain on 14 March 1939. Czechoslovakia was forced to sign away its independence the next day. Britain's spies had been busy. In July that year Polish codebreakers described to French and British agents how they had cracked the German Enigma codes. In August they handed over two of their Bomba deciphering machines, one to each country. The British Secret Intelligence Service ('MI6') received the British machine. The Second World War broke out less than a month later.

What can we observe or deduce from the interwar period? Firstly, induction can be dangerous. Where thinkers thought carefully about the evidence of the First World War, their findings were generally useful and valid. Conversely, where they took the Great War as a starting point and extrapolated from there, the results were often wrong. The wilder claims for armoured warfare, and the whole issue of long-range bombing, are good examples.

Secondly, it is wrong to say that armies simply prepare to re-fight the last battle of the last war. Some don't even do that. If they did, they would not repeat the mistakes of the last war. The French did, largely, avoid such mistakes. *Good* armies (or navies, or air forces) prepare very thoroughly to refight the last battle of the last war. *Excellent* armies think very deeply about the first battle of the next war. Unfortunately for France, that is just what the German Army had done. Put another way: the losers can learn most from a war; if they are professional. The Germans were.

The issue of institutional agendas is closely related to that. The RAF spent most of its early years defending its existence. It promoted long-range bombing in the 1930s to give it a unique and supposedly strategic role. That role challenged, and might even have threatened, the roles of the Royal Navy and the Army. Attack is often the best form of defence; institutionally the RAF promoted long-range bombing primarily to defend its own existence. We shall see how that perverse logic led to ridiculous consequences in the coming war. The US Army Air Corps, soon to become the Army Air Force and then the US Air Force, was no less guilty.

Thirdly, international affairs are often interconnected. Ostensibly Italy and Germany were strange bedfellows. Japan was an even stranger co-belligerent. As we have seen, Japan and Italy were relative newcomers on the international scene; expansionist and irredentist. Their interests happened to coincide in some ways. That was what pushed them into the Second World War. It was not necessarily because they were natural allies of National Socialist Germany. And it was their involvement that made the next war a world war, rather than a purely European conflict.

Unfinished business persists. No war resolves everything in one go. Even the Great War did not do that. Some wars resolve nothing. Some wars resolve individual issues to the extent of pushing them below the threshold of collective armed violence. Political debate may then push them below the threshold of national consciousness. But there is always unfinished business.

Related to that, wars raise new issues. The Great War created Czechoslovakia and with it the existence of the German minority in the Sudeten. The Irish War of Independence created Northern Ireland and its Republican minority. A hundred years later that minority became the largest party in the Northern Irish parliament. We shall look at the Troubles in Northern Ireland in a later chapter.

6

The Second World War: Strategy

The Second World War was the greatest in scope and scale that the World has ever seen. Yet in can be argued that in geopolitical (that is, grand strategic) consequence it was less significant than the Great War. In 1945 the United States was confirmed as the world's greatest power. The Soviet Union extended its hegemony several hundred kilometres to the west. The United Nations replaced the failed League of Nations to become a functioning international organisation. Germany and Japan were chastened for generations to come, but soon regained their position as major economic powers. To a lesser extent something similar could be said for Italy.

But no empires ceased to exist. Much of the world still consisted of European colonies. The end of the colonial system *was* clearly in sight, and the Second World War no doubt hastened that. However, as we shall see in a later chapter, the end of colonialism was in hand before 1939.

Like Chapter Two, this chapter looks at the strategy of the war. It considers the goals sought, the ways chosen, and the means applied. Under 'means' it looks at scales of effort, strategic resources and the mobilisation of national economies for war.

Before we start, however, we should consider how we know what we know: the historiography. To Anglo-Saxon audiences the history is dominated either by Winston Churchill or Dwight Eisenhower. There have been many other, excellent histories of the war; but they largely stand on what Churchill and Eisenhower wrote.

Churchill was only too aware of the 'battle of the memoirs' following the Great War. He was thrown out of government in May 1945 and immediately started to produce his own version of the facts. Churchill's book 'The Second World War' was published in six volumes from 1948 to 1953.[1] It is in practice an autobiography cunningly disguised as a history of the war. Looking back now, decades after the war, we can see considerable shortcomings and at least one glaring omission.

Roosevelt died in April 1945 and so never wrote a memoir or history. Dwight Eisenhower's 'Crusade in Europe' is clearly smaller in scope than 'The Second World War'.[2] However, it was, and is, hugely influential. That is partly because it was published as early as 1948, and partly because Eisenhower went on to be president of the USA for two terms.

Of the many books published since then, perhaps the most insightful is Chester Wilmot's 'The Struggle for Europe'.[3] It was described as 'the essential companion to the Churchill memoirs' and published in 1952. Wilmot was a BBC war reporter who interviewed many of the German commanders and politicians after the war. He also had access to a great number of German documents through his coverage of the Nuremburg War Trials. Like Eisenhower's 'Crusade', it is limited to the war in Europe. However, it seems more objective than Churchill's account. It also considers issues such as strategic raw materials, and the effects of Allied bombing, far more thoroughly than Churchill's or Eisenhower's books. Wilmot made connections between events, and between events and politics, which Churchill did not.

We look firstly at Axis strategy, and within that, Germany's. Hitler's main goal was expansion eastwards, at the expense of the Soviet Union (and Slavs generally). Hitler had some understanding of the importance of strategic raw materials. His generals generally did not. Hitler was concerned, for example, that British moves into Greece in 1941 would put the Romanian oilfields within bomber range.[4] His main strategic goal for 1942 was the Soviet oilfields in the Caspian region. Thereafter, he was repeatedly asked by his generals to shorten the strategic perimeter held by German forces. He generally refused to do so on the basis of the raw materials which a given territory contained or protected. By 1944 Germany was being attacked from four directions: the east; the west; Italy and the Balkans; and the Allied bomber offensive.

1 Winston Churchill, *The Second World War*. (New York: Houghton Mifflin 1948).
2 Dwight Eisenhower, *Crusade in Europe*. (New York: Doubleday, 1948).
3 Chester Wilmot, *The Struggle for Europe*. (London: William Collins, 1952).
4 Wilmot, *The Struggle for Europe*¸ 69.

Hitler was a poor strategist. His main method seems to have been incremental opportunism. After Austria and Czechoslovakia, he seems to have believed that he could occupy Poland without bringing Britain into the war. (Remember that it was Neville Chamberlain, who Hitler had met at Nuremburg, who declared war; not Churchill). In late 1940 he seems to have been ambivalent about invading Britain. It is certainly true that he decided to invade the Soviet Union in 1941 *before* the Battle of Britain.

German forces overran Denmark, Norway, Poland, the Low Countries and France in short order. In doing so they evicted the (second) BEF from the Continent. Consequently Britain could not conduct major counteroffensive operations in Europe for four years. However, failing to defeat Britain meant that American forces had a firm base for the invasion of Europe in 1944.

Hitler's invasion of Russia was not a strategic blunder. That idea stems from Soviet propaganda and knowledge after the fact. Germany had knocked Russia out of the war in 1917. Why should it not do so again? In 1941 the Wehrmacht (German Army) was, with its 20 or so panzer divisions, far better placed to overrun Russia than it had been in 1916 or 1917. German generals were actually well aware of Russian mud and winter: many had served there in the Great War. However, the actual plan to overrun the Soviet Union in 1941 was a bad compromise between Hitler and his generals. It had not one objective but three, and no apparent main effort.

The German surface fleet posed a considerable threat to British shipping. Its submarine fleet was initially quite small, but grew rapidly in size and effectiveness. Churchill said that it was only the U-Boat threat that really kept him awake at night.

Turning to Japan, by 1941 it already occupied much of China. Its objective on entering the war against America was to create an economic zone which would supply it with the materials (and perhaps the markets) it would need to continue to enlarge and become a major power. Like Germany, Japan had no oilfields. It also needed supplies of rubber and iron ore. The Japanese seized airfields in French Indochina in July 1941 (following France's surrender to Germany in 1940). In an escalating series of moves, America embargoed oil supplies to Japan on 1 August 1941. Japanese aircraft attacked Pearl Harbor on 7 December.

Japan had a general as its prime minister, an Emperor who was a figurehead, and major divisions between the Army and the Navy as to strategic policy. It had a brilliant, Navy-led *operational* plan. The Navy would destroy the US fleet in Pearl Harbor, and then create an outer perimeter which contained and protected the resources and markets which Japan needed. However, that was not a *strategy*. Did Japan believe, for example, that when it attacked at Pearl Harbor the Americans would simply give up? Or be incapable of mounting a counteroffensive forever?

Japan's operations to its southwest and south brought in into conflict with Dutch and British colonies, New Zealand, and Australia. Its opponents could at first do relatively little except defend doggedly. Remaining French colonies in the Far East were quickly overrun. America was ill-prepared in early 1942. The Philippines fell after five months. 100,000 US soldiers, including 20 generals, were taken prisoner. In the longer term, however, Japan was always likely to lose.

Finally, Italy. Italy had territorial ambitions (once again) in the Balkans and in east and north Africa. The latter assumed that British Egypt would fall quickly. However, in the Balkans Greece resisted fiercely. Germany had to intervene. When it did, the result was decisive: Greece was quickly overrun. That delayed the invasion of Russia by several weeks.

In North Africa, Italy suffered a major defeat almost immediately. The Wehrmacht had to intervene again. For most of the North African campaign the German contingent was just three divisions. At most, it was seven. With that German assistance, the North African campaign lasted well over two years.

Italy's forces in East Africa had been defeated largely by British Empire and Dominion troops. Italy lost, lost everywhere, and lost heavily. It lost a total of about 300,000 men killed and 1,194,000 prisoners. About half of the prisoners were interned by the Germans when Italy signed an armistice with the western Allies. At sea Italy could never really contest the Mediterranean against the Royal Navy. Without the use of the Suez Canal, Italy could not effectively support operations in East Africa.

There was little real Axis strategy. Relations between the three countries were based on mistrust and mutually inconsistent assumptions of racial superiority. Who thought they were superior to who? There was some coordination of military operations, but little else. Hitler declared war on the USA on 11 December 1942, soon after Pearl Harbor. That was probably meant to be a gesture of solidarity with Japan. It was, however, a gift to Roosevelt. Congress immediately and unanimously declared war on Germany.

Turning to the Allies, France and Britain declared war in 1939 in order to contain Germany. They could never have defended Poland directly. After the defeat of France in 1940 the French strategy that concerns us here relates to 'Free' (that is, non-Vichy) French forces. Their strategic goal, initially, was simply to liberate France.

By late 1940 the British Empire was effectively alone in facing Germany and Italy. As we have seen, Italy did not pose much of a problem. The main problem for Britain was that shipping for the Far East had to be routed around the Cape, rather than through the Mediterranean. The events of 1940 gave the German Navy access to most of the Atlantic coast of Europe.

Britain therefore developed an Atlantic defence strategy. Its goal was the defence of the United Kingdom. Britain garrisoned Iceland and earmarked up to an army corps for deployment anywhere in the north Atlantic at short notice. It significantly strengthened the garrison of Gibraltar, stationed two Royal Marine brigades in Sierra Leone and operated two air force stations in Gambia. Britain invoked the Anglo-Portuguese treaty of 1373, which remains the world's longest-standing treaty.

Put simply, Britain had to defend its empire and the sea communications to it. Hence the importance of Egypt in the mid to long term. Britain started to ship an average of 1000 men a day to Egypt. 20,000 men were *en route* at any one time.

The Japanese quickly overran Burma. British forces successfully defended the Indian border and, with it, land communications to China. By doing so, it could supply Nationalist Chinese forces fighting the Japanese. Churchill was well aware that in order to defeat Germany, America would have to enter the war. When it did, British strategy became largely synonymous with Allied strategy, described below.

Defence of the British mainland included the Battle of Britain and then the Battle of the Atlantic. The former was a major air defence operation; the latter a multi-year antisubmarine campaign which peaked in 1943. After the battle of Britain, Britain's main offensive activity was bombing Germany. In 1942 the US Army Air Force (USAAF) joined the RAF in what became the Combined Bomber Offensive (CBO).

Allied *grand* strategy was formalised through a series of conferences. Churchill and Roosevelt conferred six times, in North America and at Casablanca. They then met Stalin at Teheran and Yalta. The latter stages of the war were then discussed at Potsdam in July 1945. By that time Roosevelt was dead. Churchill was out of office (he was replaced by Clement Attlee, although he did attend the Potsdam conference.) Roosevelt had been replaced by Harry Truman, his vice president. Truman was probably more accommodating towards Stalin than Roosevelt would have been. (Churchill had not met Truman before Potsdam.)

The origins of the United Nations lay in political discussions in London in 1941. Those discussions laid down principles that would lead to the Atlantic Charter in December 1941, and then the 'Declaration by the United Nations' of 1 January 1942. Importantly, the London discussions included an agreement by 13 governments (several of them in exile) and the Free French. They would make no separate peace with the Axis powers.

Allied *military* strategy was determined by the Combined Chiefs of Staff (CCS), a standing body created in January 1942. It represented the British Chiefs of Staff Committee and the US Joint Chiefs of Staff. (The latter sometimes attended in person). The CCS was a remarkably effective Allied institution.

North Africa had to be cleared in order to defeat Italy and bring American forces into Europe. Sicily and Italy then had to be invaded, for several reasons. Doing so neutralised Italy: Italy agreed terms the moment that Sicily was invaded. That opened the southern Mediterranean to Allied shipping. Invading Italy brought southern Germany within bomber range and brought significant French forces into the war from north Africa. Later, it kept some German forces away from Normandy. It also showed Stalin that the western Allies were actively involved in fighting. Subsequent American and French landings in southern France in August 1944 (under Operation Dragoon) opened up further sea lanes of communication. They also allowed French forces to complete the liberation of France.

The aim of Operation Overlord, which became the Normandy Campaign, was to gain a lodgement on the continent of Europe from which operations could be developed against Germany.[5] As we shall see

5 Stephen Brooks ed, *Montgomery and the Battle of Normandy: A Selection from the Diaries, Correspondence and Other Papers of Field Marshal the Viscount Montgomery of Alamein, January to August 1944*. (Stroud: The History Press, 2008), 69. This work is essentially the unedited version of Montgomery's personal diary.

in Chapter 7, it was brilliantly successful. Subsequent operations against Germany had not been considered in detail. Furthermore, little or no consensus exists even today as to how the Western allies defeated the German Army in the west. That will also be considered in Chapter 7.

America and Britain initiated the CBO at the Casablanca conference. In practice the CBO became an end in itself, as we shall see. Between 1941 and 1943 Churchill and the RAF had persisted with bombing Germany despite realising that it was having 'far less than decisive effect'.[6] His reason was that, politically, he needed it as a substitute for the missing 'second front' (the Eastern Front being the first). Much of the purpose of the Allied bombing campaign was, at least at first, political signalling. Churchill had been signalling to the United States, the Soviet Union and the British population.

Soviet strategy was essentially simple. It was to resist the German attack, then counterattack to defeat German and establish a buffer zone in eastern Europe. For us the real interest in Soviet military operations lies at the operational level. That is described in Chapter 7.

The defeat of Japan involved three separate theatres: the central Pacific, the southern Pacific and southeast Asia. The first centred on the American-led liberation of the Philippines. The second was a truly amphibious campaign on a vast scale. It brought American forces within bomber range of Japan, and then to the point where Japan could be invaded. The largely British campaign in southeast Asia liberated Burma and then other occupied territories. It also tied down several Japanese divisions. It has been questioned, however, whether American strategy in the Pacific was coherent.[7] Did the Army-led campaign in the Philippines contribute to the defeat of Japan? Did it support the amphibious campaign? Or vice versa? Was the Philippines campaign even necessary?

Japan surrendered soon after two atomic bombs were dropped on Japanese cities. That demonstrates nothing more than overwhelming, one-sided technological advantage. But did the atomic bombs end the war? The atomic bombs were dropped on 6 and 9 August 1945. The Soviet Union declared war on Japan on 8 August. Japan surrendered on 15 August. The

6 Colin S Gray, *Airpower for Strategic Effect.* (Maxwell, Alabama: Air University Press, 2008), 50.
7 Gray, *Airpower for Strategic Effect*, 65.

Japanese government had previously considered that it could not defend its homeland against an attack from the Soviet Union. That might lead to the Soviets occupying large parts of Japan. Occupation by the USA would be preferable. So there is good reason to doubt that it was the atomic bombs that prompted Japan's surrender.

We now turn to the strategic means, that is the national effort, applied. The Second World War was vast. Figure 6-1 gives an indication of the number of divisions raised and employed by the major powers.

Nation	Divisions Raised and Deployed	Peak Manpower; or	Manpower at End of War; or	Total Manpower Mobilised
Australia	7			727,200
Belgium	22		650,000	
Bulgaria	15 (?)			1,011,000
Canada	5		474,000	
China	?			14,000,000
Finland	16		270,000	
France	96		437,000	
Germany	414	6,500,000		
Greece	20	540,000		
Hungary	32		210,000	
India	17		2,100,000	
Italy	99	2,563,000		
Japan	110		5,500,000	
Netherlands	8			400,000
New Zealand	2		157,000	
Norway	6			90,000 (?)
Poland	41			2,400,000 (?)
Romania	31	1,225,000		
South Africa	3			208,000
United Kingdom	34			3,778,000
United States	94		5,851,000 & 456,000 USMC	
USSR	680		6,000,00	

Figure 6-1 Second World War Land Forces

Class	RN	France	Germany	Italy	USN	IJN
Battleships and Battlecruisers	15+5	7+4	5+11(a)	6+?	17+15	10+3
Aircraft Carriers	7+6	1+2	0+2 (b)	0+2 (b)	7+11	12+7
Cruisers	66+23	19+3	8+9 (c)	19+?	41+42	38+6
Destroyers	184+52	78+27	22+12	59+?	171+188	126+43
Frigates and Corvettes (d)	45+56					
Submarines	60+9	81+38	57+?	116+?	114+79	68+?

Figure 6-2 Fleet Size at the Outbreak of War
Notes:

(a) Only 2 (Bismarck and Tirpitz) completed. (b) None completed. (c) Only 1 completed. (d) ie, purpose-built maritime convoy escorts.

Similarly Figure 6-2 shows fleet sizes at the outbreak of war. Numbers shown are 'in commission' + 'in building.' [sic]. Figure 6-3 shows British and American naval construction and losses.

Type:	Britain:		America:	
	Built:	Lost:	Built:	Lost:
Battleship	5	5	12	2
Aircraft Carrier	52	8	98	11
Cruiser	39	30	87	10
Destroyer	120	110	368	71
Convoy Escort	473	58	578 (a)	10
Submarine	152	77	87	53

Figure 6-3 British and American Construction and Losses
Note: (a) Of which, 108 for Britain.

Put simply, Britain narrowly replaced its losses of battleships, cruisers and destroyers. That apart, Britain and America built far more ships than they lost. As another indication, Germany had 57 U-Boats at the outbreak of war. It built 871 new boats. It lost 785.

Aircraft production was similarly massive. Germany produced 39,807 aircraft in 1944 alone. Japan produced 28,180. However the Allies produced 163,035. In April 1945 Bomber Command alone had 1,609 aircraft with crews. Almost all were heavy bombers. Altogether the Western Allies deployed almost 47,000. At its peak the Luftwaffe had about 5,000 aircraft at front line.

The proportion of national resources allocated to bombing was probably the biggest British strategic mistake of the war. 1,573,000 civilians were employed building aircraft, most of them heavy bombers. At its peak in June and July 1943 the Air Ministry and Ministry of Aircraft Production had 3,619,500 civilians working for them, which was 20 per cent of the entire national workforce. The RAF grew to 1,012,000 men and women.

We should not overlook German missile bombardment. 9,521 V-1 flying bombs and 1,402 V-2 ballistic missiles were fired at Britain from the summer of 1944. A further 2,448 V-1s and 1,664 V-2s were fired at Antwerp after it was liberated in September. Other than loss of life and damage to property, the bombardment of Antwerp had little effect. Its port was critical to Allied logistics, once it was opened in November 1944. However, the port remained open despite the bombardment. That was in large part due to Allied air defences. The cost of the V-bombs to the German war economy was considerable.

The mobilisation of the home fronts underpinned such massive war production. Nothing on such a scale has ever been seen before or since. However, the scale and scope of that effort varied considerably. The entire US automobile industry was switched to war production. Not least, it produced 642,000 jeeps. However, beyond that, the American population was directly affected by the war relatively little. In Britain practically every man and women of working age was serving or involved either in productive work or childrearing. German women were not conscripted nor directed. Neither were Japanese women. However, millions of both undertook war work, typically on farms or in factories. In the Soviet Union practically everyone was involved in the war effort. Nevertheless, even there there were limits to the extent to which the authorities could coerce industrial and agricultural workers.

The effort was generally significant, but was it effective? As a broad generalisation the Western Allies mobilised their economies more effectively than the Axis powers. Allied advantages in cryptology, radar and atomic weapons indicate that, on balance, America and Britain engaged their scientists and engineers better than Germany and Japan did.

Size matters. In 2024 dollar terms, in 1938 the GDPs of the three Axis nations totalled $1.58 trillion dollars. The GDP of Britain, its Empire and Dominions totalled $1.64 trillion. America's totalled $1.93 trillion, and the Soviet Union's $865 billion. Thus the Allied total was over $4.435 trillion. Assuming even reasonably efficient mobilisation, the Allies would win a long war. After America entered the war Churchill wrote '[s]o we had won after all!'[8]

What can we observe or deduce from the strategy of the Second World War? Firstly, we should consider it, at least in part, as a war of resources. We shall look at how that worked in practice in the next chapter. We can say here, however, that the war was not won by bombing. Bombing did not end German war production, nor break German morale. In practice Italy surrendered just before it was invaded. Germany was overrun. Japan surrendered shortly before it was invaded. Japan had, in effect, already lost. As we shall see, bombing *contributed* in all three cases (and the two atomic devices were, of course, delivered by bombing).

Overall the outcome of the Second World War was a result of effective Alliance strategy. Grand strategy (developed through major conferences) and military strategy (delivered principally by the CCS, for the western Allies) provided the proper application of overwhelming force, as Churchill put it. Strategy matters. Having effective institutions with which to make it is critical.

More than three generations later, Germany and Japan show no signs of returning to war. No political issue is ever resolved for all time. However the unconditional surrenders imposed on those two countries probably comes as close to such a resolution as there has ever been. That is, as far as we can know (since we cannot know the future).

So: if you wish to impose a genuinely lasting political solution on another country, first impose an unconditional surrender; and then enforce

8 Churchill, *The Second World War*. Abridged one-volume edition. (London: Cassell and Co Ltd, 1959), 492. Note the tense.

it. Note the verb 'enforce'. If you do not do that, do not be surprised if the political resolution to the conflict does not endure. Lastly, however, it was not 'total war'. Nations tend to mobilise to the extent that they believe that the situation warrants. The Second World War was vast. The effort was immense. Some nations suffered grievously. But the effort was not total. German women were not conscripted. In Britain bread was not even rationed until *after* the war. The idea of 'total war' is overstated. It is mostly something for political scientists and historians to argue about.

7

The Second World War: Campaigns and Operations

We start with the same caveat as for Chapter 3. The operational (or campaign and theatre) level was not really recognised during the Second World War. The Red Army, the Wehrmacht and the Luftwaffe were broadly aware of it. British and American doctrine were moving towards the identification of the campaign as a discrete level of operations but, as we shall see, there was no deep understanding of the identification of ends, ways and means at the campaign level which we can identify today.

To an Anglo-Saxon audience, much of our knowledge of events comes from the memoirs of Patton, Eisenhower, Bradley and Montgomery. Importantly, Montgomery's diaries were not fully available until 2008 or so.[1] The Allied commanders' accounts were overlain in the 1950s and 1960s by those of German generals such as Guderian and von Manstein. From the 1970s western 'Sovietologists' worked from official Russian accounts. They provided useful material and insight. However, in a number of areas Soviet sources either concealed or lied. That was partially rectified when the Soviet archives were opened briefly in the early 1990s. Many misperceptions persist.

We cannot examine every campaign here. Chapter 7 looks at some naval campaigns, some aspects of aerial warfare, some land campaigns, and then some aspects of irregular warfare.

We will consider three naval campaigns. They are the war in the Mediterranean against Italy, the neutralisation of the German surface fleet, and the antisubmarine aspects of the Battle of the Atlantic. Strategically,

1 Brooks, *Montgomery and the Battle of Normandy*. To repeat: this book is essentially the unedited version of Montgomery's personal diary.

from mid-1940 the Royal Navy (RN) faced two opponents: the Germans in the North Atlantic and the Italians in the Mediterranean. The RN was larger than both combined. Possession of Gibraltar allowed it to move ships between the two theatres and maintain an advantage in both.

In 1940 Malta was a British colony. It lies midway between Gibraltar and Egypt. Malta was both a relatively safe anchorage for the RN and an unsinkable base for aircraft. Italy and Germany wanted to transit the Mediterranean north to south (and vice versa) to supply their armies in North Africa. Britain introduced heavily-escorted convoys to supply and reinforce Malta from the east and the west. The Italian air force and navy, together with the Luftwaffe, attacked the convoys. The British lost 31 merchant ships from 35 convoys up to August 1942.

The RN combined fleet actions with its convoy operations. On 11-12 November 1940 an antishipping strike by carrier aircraft against the Italian fleet at anchor at Taranto was highly successful. Four Italian warships were sunk. The rest of the Italian fleet withdrew north to Naples. The attack at Taranto informed and inspired the Japanese attack at Pearl Harbor a year later. British submarines from Malta sank 185 ships in the next 24 months or so.

The Allies landed in Morocco and Tunisia in November 1942 (in Operation Torch). That, and the battle of El Alamein, allowed Allied aircraft to operate further and further into the central Mediterranean. More airfields were built on Malta, allowing far more aircraft – both fighters and anti-shipping types – to operate from there. The last major convoy reached Malta in August 1942 under Operation Pedestal. Malta-based air and naval forces then sank 230 more ships in the next 164 days.

During the war Italy had, or obtained, 983 merchant ships. 565 were sunk or captured before July 1943, when Italy left the Axis. It is clear that the RN and RAF fought a major, effective, and largely unannounced anti-shipping campaign. It is not obvious why that is not more widely known.

British naval and air forces did not stop the Axis deploying forces to North Africa, nor supplying them. They did cause significant disruption. British submarines sank two troopships, and in doing so killed more Germans soldiers than died at El Alamein (over 5,000). Allied air and naval forces did, however, prevent the evacuation of Axis forces from Tunisia in May 1943. 275,000 men, including over 100,000 Germans, surrendered.

Turning to the Atlantic, Germany had a total of 11 major warships at the beginning of the war. There were four battleships, three 'pocket' battleships and four heavy cruisers. No more were built. As in the Great War, they were a major threat to shipping. In an echo of 1914, the pocket battleship *Graf Spee* was operating in the south Atlantic in late 1939. She was heavily damaged by three British cruisers and scuttled. The heavy cruiser *Blücher* was sunk by a coastal battery in Oslo Fiord in April 1940. In Operation Berlin, from January to March 1941, the battleships *Scharnhorst* and *Gneisenau* sank 22 merchantmen. Two battleships, the *Bismarck* and the *Scharnhorst,* were then sunk in fleet actions in May 1941 and December 1943.

After *Bismarck* was sunk, seven of the remaining eight ships were damaged by torpedoes from British submarines. They all required lengthy repairs. When ordered to sea in December 1942, *Hipper* and *Lützow* failed to cause much damage. All German warships were then ordered back to port. They rarely operated in the Atlantic again. Construction of new ships was stopped. The Allies did not know that. Considerable Allied forces, including battleships and aircraft carriers, were committed to guard against the threat posed by the remaining German ships. Four were sunk in harbour by the RAF, but only after November 1944. By then the RAF had effectively run out of land targets to bomb. The last two German warships survived.

Strategically, control of the Atlantic was important to the Allies for two reasons. The supply of food and war material allowed Britain to stay in the war. Control of the Atlantic also allowed Canadian and American forces to reach Europe and, particularly, Britain. German submarines significantly interrupted the former, but not the latter. Graphs show the progress of the German submarine campaign quite clearly. Figure 7-1 shows Allied shipping losses by quarter.

Losses fell in late 1941, mostly due to convoying and simply building enough escorts. Losses rose dramatically in 1942 as U-boats attacked unprotected American shipping in the western Atlantic. They fell over the next year as the US Navy introduced convoying. By October 1943 losses were relatively small. Figure 7-2 shows U-boat losses over the same period.

Losses of U-boats rose significantly from mid-1942 to mid-1943, but from a very low level. However, Figure 7-3 shows U-boats launched, sunk and available against merchant ships sunk.

After 1942 there were plenty of U-boats: but they simply couldn't sink merchant ships. Note that from the second quarter of 1943 U-boats

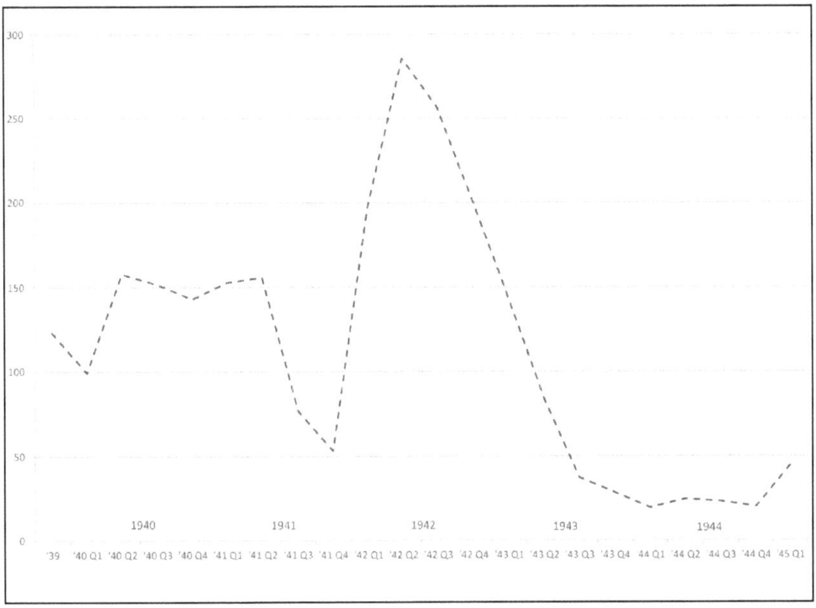

Figure 7-1. Allied Merchant Shipping Losses by Quarter

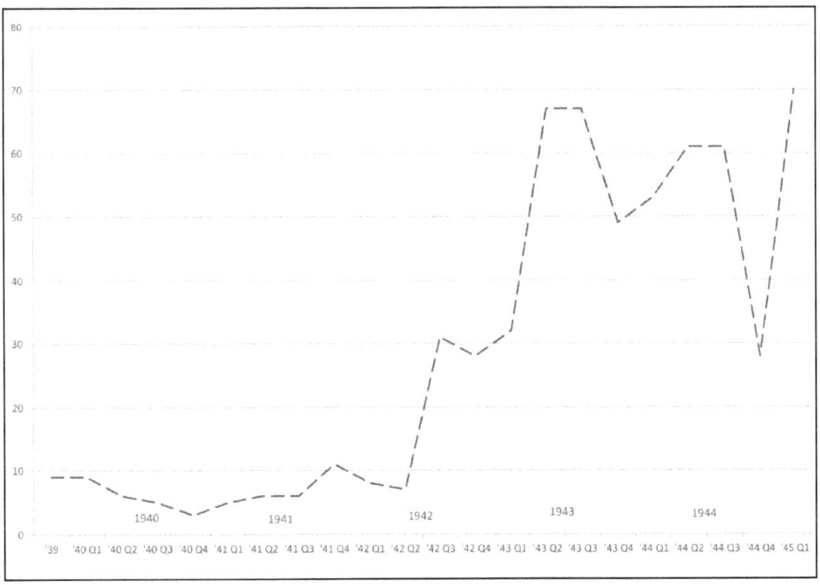

Figure 7-2. U-Boat Losses, by Quarter

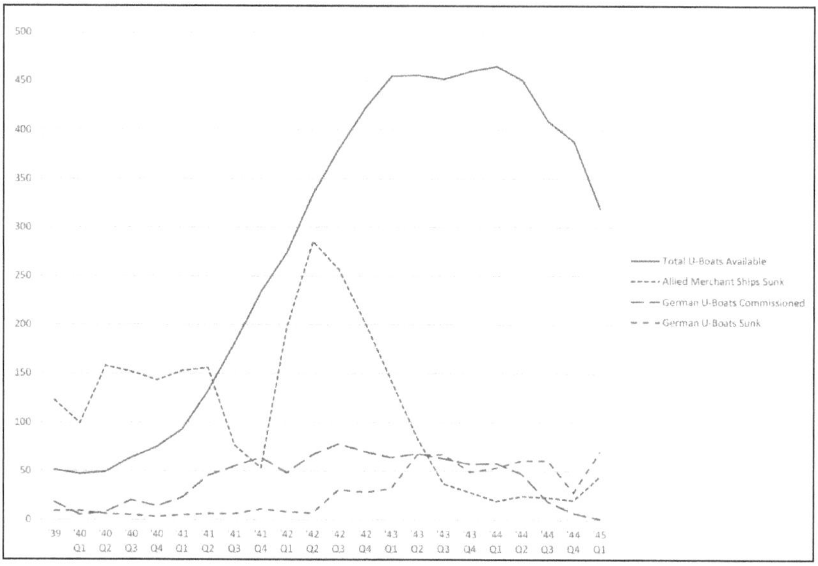

Figure 7-3. U-Boats Launched, Sunk and Available, by Quarter

were being launched and sunk in roughly equal numbers. However, more U-boats were being sunk than allied merchant ships.

Better sensors, such as airborne radar, helped. So did better weapons, such as forward-firing salvo depth-charge launchers on ships. Aircraft, either shore-based or from escort carriers, were highly important. Aircraft sank more than half of all U-boats. To do so they had to spot the U-boat on the surface. Patrolling submarine deployment routes, and searching ahead of convoys, forced the U-boats to stay submerged for longer, or to surface only at night. That made the submarines slower and less effective. There was also a real-time battle to break naval codes, and develop better tactics, in order to get ahead in the battle between the submarines and the convoy escorts. Deterrence, and better intelligence, saved ships; not sinking U-boats. The human cost was significant. The German submarine service lost about 28,000 men killed, or about 75 per cent of all those who put to sea. About 72,000 Allied seamen died.

Troop convoys were provided with twice as many escorts as merchant shipping. Large passenger liners were also converted into troopships. Some could carry over 15,000 men per sailing and sailed almost unescorted. At over 35 knots, no submarine could conceivably catch them. Incredibly few servicemen were lost crossing the Atlantic by ship.

The commitment of the RAF to the Battle of the Atlantic was limited. The key asset was long-range, long-endurance, four-engined bombers. Aircraft such as the Very Long Range ('VLR') Liberator could cover the whole of the north Atlantic from bases in Iceland or Northern Ireland. However for most of the war no more than 50 RAF VLR aircraft were made available. That contrasts with attacking Cologne with 1,000 bombers on one night in May 1942.

We now turn to war in the air. We shall look briefly at the air war in 1940. We will then look at two aspects of Allied long-range bombing. The first is what direct results it had. The second is what else it achieved. Finally, we will consider some other facets of the Allies' use of air forces, particularly in northwest Europe.

When looking at the events of the summer and autumn 1940 the first thing to notice is a great silence. In 1940 Bomber Command had four Groups of medium bombers. What did they do before the surrender of France in June 1940? We find reference to dropping leaflets over German cities. We then find reference to one raid after the German invasion of the Low Countries but apart from that, very little. We can be fairly sure that they did something, but what? A detailed search reveals that those four bomber groups made just 11 attacks in those two months. Overall, 'in the early months, and even years, of World War Two, Bomber Command was unfit for purpose.'[2]

The Luftwaffe did conceive of its attacks on Britain in mid-late 1940 as an 'air operation' in the operational sense. Its primary task was 'to eliminate the English [sic] Air Force both as a fighting force and in its ground organisation.'[3] It was then to 'strangle England's supply lines by destroying her ports and her shipping'. Those two phases were to be augmented by bombing aircraft factories at night.

A combination of poor intelligence, overoptimism, British radar, superior RAF ground-air communications and a brave and well-conducted fighter defence defeated the daylight phases. The Luftwaffe could not continue bombing by day. However bombing cities at night, hoping to hit aircraft factories, continued. The British called it 'the Blitz'. It continued until the spring of 1941. The Luftwaffe then redeployed its air fleets for the invasion of Russia.

The RAF also discovered that it could not sustain daylight long-range bombing. There appears to be some consensus that the subsequent

2 Gray, *Airpower for Strategic Effect*, 106.
3 Wilmot, *The Struggle for Europe*, 32.

long-range 'strategic' bombing had a major effect on the outcome of the War. It did not. To justify that, firstly (and to repeat): long-range bombing is not 'strategic'. Damage caused by weapons and munitions is tactical. A series of attacks can be called an operation or campaign, and can have a theatre- or campaign-level impact. Clearly any campaign *should* have a strategic impact. That is, it should affect the outcome of the war. Importantly, was that the case with bombing in the Second World War?

The answer is 'yes'; but also 'to some extent'; and 'much less than people think'. Firstly, German and Japanese civil morale did not crack. That is despite horrific amounts of bombing and hundreds of thousands of deaths. What was the effect on industrial production? Our main evidence comes from the United States Strategic Bombing Survey of September 1945.[4] It did a reasonably good job of answering the question 'what did Allied bombing do?' What we need, however, is the answer to the question of 'what happened to German war industries?'

The German (and Japanese) war economies depended principally on coal, oil, and iron ore with which to make steel. Germany had three main industrial areas: the Ruhr, in the west; Saxony, in central Germany: and Silesia, in the east (modern Poland). The Ruhr was the most important. However (for example) most of Germany's synthetic oil production was located in the Saxon coalfields.

Germany was never going to run out of coal. It gained even more when it overran the Donetz Basin ('Donbas') in Ukraine in 1941. In doing so Germany also gained a lot of iron ore, albeit temporarily. (The Red Army liberated the Donbas in August 1943). Iron ore was, however, a problem. From 1933 to 1943 Germany imported 43 per cent of its ore from Sweden. Swedish ore produced more steel per ton, and the steel was of higher (weapons-grade) quality. American diplomats persuaded Swedish insurance companies to stop insuring ships carrying iron ore to Germany in 1944. The last shipment arrived in Germany that September.

Germany received about a quarter of its oil from Ploieşti, and some from Hungary. Despite heavy bombing, Ploieşti *exported* as much oil per month to Germany in 1944 as it had in 1942. That is, until August; when the Romanian army liberated the Ploieşti oilfields. The output was diverted to

4 United States Strategic Bombing Survey: Summary Reports (European War and Pacific War) (af.edu) accessed 10.01hrs GMT 14 February 2024.

Year	Month	Bombs Dropped[1] (Jan 42 = 100)	Military Production[2] (Jan 42 = 100)	Tank Production[2] (Jan 42 = 100)
1943	Jan	215	182	154
	Feb	503	207	169
	Mar	527	216	210
	Apr	538	215	289
	May	648	232	465
	Jun	768	226	340
	Jul	886	229	367
	Aug	1,054	224	328
	Sep	970	234	495
	Oct	839	242	454
	Nov	1,119	231	364
	Dec	1,318	222	415
1944	Jan	1,746	241	438
	Feb	1,539	231	480
	Mar	2,520	270	498
	Apr	2,455	274	527
	May	3,452	285	567
	Jun	5,148	297	580
	Jul	4,789	322	589
	Aug	5,112	297	558
	Sep	4,810	301	527
	Oct	5,098	273	516
	Nov	4,658	268	571
	Dec	4,598	263	598
1945	Jan	3,272	227	557

Figure 7-4. Germany: Bombing Effort and Effect, 1943-5

1 Ellis, John, *The World War Two Databook*, (London: Aurum, 1993), Table 22.
2 Tooze, Adam, *The Wages of Destruction. The Making and Breaking of the Nazi War Economy*, (London: Allen Lane, 2006), Table A6.

the Soviet Union. The amount of oil that Germany had *received* from Ploieşti, however, had fallen heavily. That was largely due to Allied air interdiction. Much of that resulted from attacks on oil barges on the Danube.

Germany had started making synthetic oil from coal in 1934. The main factory, the Leuna works, was in Saxony. Persistent bombing slowly reduced its output throughout 1944 (although the number of workers killed was astonishingly low). Germany was building three new plants. They were in Silesia, further away from the bomber bases in Britain.

However, the Red Army overran Silesia in January 1944. The German armaments minister Albert Speer described that as 'unbearable'.[5] That is, the German economy simply could not bear the loss of Silesian industry. Not least, it lost its potential alternative sources of synthetic fuel. As late as March 1945 German forces conducted a counteroffensive in southern Hungary in an attempt to secure the last source of foreign oil.

Air Chief Marshal Sir Arthur ('Bomber') Harris took command of the British Bomber Command in February 1942. Allied bombing effort increased by 5,112 per cent. Yet German overall industrial production *increased*, threefold (see Figure 7-4). German tank production peaked in December 1944 at *six times* its 1942 level.

In summary: oil from Ploieşti stopped arriving in August 1944. Swedish iron ore deliveries stopped in September. German tank production peaked that December. 'Unbearably', Silesia was overrun in January 1945. *None* of that was due to long-range bombing. Did bombing have an effect on German war industry? Of course it did. It was hugely *disruptive*. Not least, it severely disrupted transport, including the movement of Romanian oil to Germany. However, as the Allies had discovered early in the war, German industry became increasingly efficient at re-establishing (and even increasing) production.

Factories could be back in production in a few days. Production became more resilient through diversification and duplication. Production increased. That was quite possibly *because* of the bombing; and certainly *despite* the bombing. At best, two claims can be made in support of the effect of long-range bombing on German industry. The first is that production would have increased even further without it. That is unlikely. By 1944 the

5 Wilmot, The Struggle for Europe, 698.

German economy was running white-hot. It was running critically short of other assets such as labour (despite the use of millions of slave labourers).

The second, weaker claim is that German resources had to be diverted to defend against the bombers. The main answer to that is that if the bombers' objective was diversion of resources, it could have been achieved (by air forces) better and at less cost. British scientists told ministers that at the time. The second answer is that the diversion of some resources (such as guns and gunners for antiaircraft defence) does not stand up to analysis. It was, numerically, not significant.[6] Furthermore, the previous criticism applies: any such diversion of resources could have been achieved better and at less cost.

At first, Britain largely bombed Germany for lack of anything better to do. From 1940 to 1944 it could not attack Germany directly on land. Unfortunately long-range bombing had become an institutional imperative for Allied air forces. It was, as we have seen, important political signalling. Once again, that signalling could have been achieved better and at less cost.

'Bomber' Harris was not a nice man. He threatened to resign often enough to know that he would not be sacked. He could then largely do what he wanted. He was remarkably single-minded, which can be a good thing. The problem lay with the strategic direction and resources allocated. British, and American, long-range bombing was disproportionate. It was disproportionate to the effect achieved, which (admittedly) was hard to know in advance. It was disproportionate to the evidence as it was gathered, which was then deliberately and systematically ignored or argued away. Finally, it was disproportionate to the objectives which a more dispassionate assessment would have indicated (such as disruption, diversion of resources and political signalling).

Churchill endorsed the initial bombing of Germany. He personally ordered the first attack on Berlin. He then supported the considerable growth in resources allocated to Bomber Command. At that point his memoirs go silent. He almost never mentioned 'strategic' bombing again.

6 Only about 100,000 adult males served in the relevant air defence units, and German armies did not run short of guns.

Churchill visited Berlin for the Potsdam conference in July 1945. Berlin was totally wrecked. At the time very few buildings in the city, across several square miles, even had a roof.[7] Churchill would have seen that.

American bombing of Japanese cities was, if anything, even worse. Much of the cities were built of wood and even paper. US air forces deliberately copied British firebombing tactics in order to create firestorms. They burned over 300,000 Japanese civilians to death. The main impact was pictures of burnt-out cities on the front pages of newspapers in the USA. That reinforced the case for an independent air force.[8] Air force commanders appear to have decided to firebomb Japanese cities primarily to further their domestic political objectives; a process described as 'egregious'.[9]

In Europe, US Air Forces generally bombed in daylight. They initially took grievous losses, but persisted. The breakthrough came with the development of long-range fighters (primarily the re-engined P51 Mustang) which could escort bombers all the way to Berlin. Waves of fighters then bludgeoned the Luftwaffe day fighter force almost to extinction. Attacking ground transport (which the Allied bombers did, after considerable prompting) reduced the Luftwaffe's fuel supply. The USAAF achieved a high degree of air superiority over Europe, and particularly France, before D-Day. That was a major operational achievement. Allied air forces then actively maintained that superiority until the end of the war.

There were many benefits. Allied land forces benefited enormously from air interdiction and close air support. They could, and did, move freely. Airborne forces could be inserted, with considerable safety. Some logistics could be delivered. That included dropping supplies to special forces and resistance groups. The evacuation of casualties to the UK by air was fast, life-saving and life-changing, and almost completely safe.

However, according to one American airpower historian, '[t]he heavy bombers were used in Europe beyond the time their employment could be justified strategically for no better reason than that they existed.'[10] Air

7 Several minutes of cine film taken from an aircraft flying over Berlin in May 1945 shows almost no intact roofs. Perhaps 30 per cent of the buildings are rubble. Virtually all were damaged. Video Berlin - Aerial Views Of Damage (1945) (youtube.com) accessed 10.22hrs GMT 14 February 2024.

8 William W Ralph, 'Improvised Destruction: Arnold, LeMay, and the Firebombing of Japan', *War in History* Vol 13 No 4, 495-522.

9 Ralph, 521,

10 Gray, Airpower for Strategic Effect, 143.

forces generally avoided public discussion of bombing policy and resorted to euphemism when announcing results. The RAF had a 'dehousing' policy which actually meant deliberately targeting civilians. The USAF used the term 'non-precision bombing' to refer to the area bombing of German towns and cities. It referred to attacks on railway 'marshalling yards'. Conveniently, that concealed the fact that, in Europe, the great majority of railway sidings were located in the middle of cities.

Air interdiction did not stop V1 and V2 launches. Air intelligence did not detect the buildup of forces for the German Ardennes offensive, the biggest German operation in the west after D-Day. Interdiction did not prevent that buildup, nor the constant movement of German armoured forces by rail. It slowed the movement armour to Normandy to some extent, discussed below. However, the main issue there was the limited capacity of the French railway network.[11]

Overall, air forces had little *direct* strategic impact. In Europe and elsewhere they made a major contribution to the success of land campaigns and as both the Luftwaffe and the RAF showed in the Battle of Britain, the use of air forces does need to be considered at the operational level. That is, the theatre and campaign level.

We shall look at land campaigns in four theatres: France and the Low Countries in 1940; North Africa; the Eastern Front; and northwest Europe (from D-Day).

We can make two major (and one minor) observations about the Fall of France at the operational level. Firstly, major armoured encirclements work. France fell as a result of a major envelopment (against the English Channel), followed by an exploitation in which German armoured columns cut up the remaining French defences. France was defeated for the loss of just 43,000 German dead. Compare that with 1.7 million dead, to *fail* to defeat France, 20 years earlier.

Secondly, the French Army made a major and typically unnoticed error. It committed its operational reserve, the Ninth Army, to Holland as soon as the Wehrmacht attacked on 10 May. *It did not then create a further reserve*. When Churchill discovered that, on 16 May, he was horrified: with good reason. There was no French operational reserve when the

11 Niklas Zetterling, Normandy 1944. German Military Organization, Combat Power and Operational Effectiveness. (Oxford: Casemate Publishers, 2019), 95.

Wehrmacht, and particularly Guderian, had crossed the Meuse on 13 May. Historians generally overlook that.

The minor but important point concerns the historiography. Most of the accounts describe Guderian's XIX Panzer Corps leading the dash from Sedan to the Channel. They ascribe the French collapse to that. That is largely a product of the literature, and particularly Guderian's book. The *envelopment* was completed on 20 May when XIX Corps reached the Channel (more precisely, the mouth of the River Somme at Abbeville). However the *collapse* is better ascribed to French panic at reports of panzers approaching Rethel and Laon on 15 May. The French prime minister, Paul Reynaud, was in shock. That is five days before Abbeville, but just two days after the Wehrmacht crossed the Meuse. The leading formation was XXXXI Panzer Corps and particularly 6th Panzer Division. It wasn't Guderian. Nor was it Rommel. Historians have not checked their maps.

We will consider North Africa, and particularly the Battle of El Alamein, in some detail in Chapter 8. Here we will look briefly at some of the operational aspects. Firstly it should be remembered that, for the great majority of the campaign, the British were outnumbered. Even at El Alamein 10 British divisions attacked 14 German and Italian divisions. Montgomery took over command of Eighth Army on 13 August 1942. He was determined not to attack Rommel at El Alamein until he was ready. He resisted repeated urgings from Churchill. The one imperative was to attack, and if possible to win, before the Allied 'Torch' Landings on 8 November. Montgomery attacked on 23 October. Rommel withdrew on 3 November.

What followed was one of the longest pursuits in history. The Afrika Korps successfully reached the Mareth Line, on the border of Tunisia, 104 days later. The Mareth line is about 2,100 kilometres from El Alamein. Montgomery had tried to envelop the Germans six times. On each occasion Rommel had halted, gave battle, and slipped away. Montgomery made some logistic mistakes, but had never had time to reorganise his logistics for a long pursuit. So logistics did hamper Montgomery's ability to catch and defeat Rommel. However, more broadly, the operation is a good example of the general superiority of the withdrawal over the pursuit, over long distances and long periods.

We should briefly mention the fighting at Cassino in Italy, from December 1943 to June 1944. We tend to focus on four battles at Cassino and an amphibious landing at Anzio. Look at it another way. To the

Germans, that was a successful operational defence for more than five months. Additionally, General Mark Clark (commanding the Fifth US Army) deliberately and wilfully turned his VI Corps north from Anzio, in order to liberate Rome the day before D-Day in Normandy. That gained a brief moment of fame for Clark in the American press. In doing so Clark chose to not encircle (and hence destroy) the German Tenth Army, east and south of Anzio. Allied armies then had to fight the Tenth Army north of Rome for the rest of the war. Clark's action was disobedient. If Clark had been British, Alexander would have sacked him. If Alexander had been American, he would have sacked Clark. Is a moment's glory worth forcgoing the destruction of a whole German Army? Conversely it should be said that the final attacks from Cassino, and from Anzio, and were both something approaching operational-level breakouts.

We can make several observations about the Eastern Front. In 1941, as we have seen, the Wehrmacht had three major objectives. They killed and captured huge numbers of Russians but achieved none of their operational objectives. It is reasonable to believe that they might have taken Moscow, and decisively beaten the Red Army trying to defend it, if they had focussed on that.

In 1942 the Wehrmacht might have captured the Caspian oilfields if Hitler had not been sidetracked by Stalingrad. That, however, would not of itself have defeated the Red Army.

The Red Army conducted *two* major operations after Stalingrad. In Operation Uranus it destroyed the German Sixth Army in an armoured encirclement. However, in the subsequent exploitation up they lost 86,000 men when Manstein counterattacked at Kharkov. The Germans lost 11,500.

The other major Soviet offensive was Operation Mars, west of Moscow. General, later Marshal, Georgiy Zhukhov commanded it. It was a total failure. The Red Army lost 335,000 men. The Soviets simply airbrushed it from history.

This is not the place to discuss whether German generals tried to absolve themselves of involvement in atrocities on the Eastern Front: the so called 'clean hands' myth. Like most moral questions the issue is far from clean-cut, which several writers seem to ignore. Some of those writers go much too far, however, when they go on to suggest that we can learn nothing from the Wehrmacht's operations on the Eastern Front, because everything they did is inevitably tainted. That is ridiculous. It shows a lack of intellectual detachment.

What happened at Kursk in 1943? The Germans assembled 14 of their 20 or so panzer divisions and attacked Soviet defences prepared in great depth. We now know that the Germans were tactically highly successful and that the Soviets took massive losses. However, operationally, the Wehrmacht ran out of *infantry* divisions. As their panzers advanced, they created long flanks. Lacking infantry divisions, the panzer divisions had to protect their own flanks. So they became progressively weaker at the critical point. The operational and then the strategic initiative passed to the Soviets.

The following year, in Operation Bagration, the Red Army effectively destroyed the German Army Group Centre. How? Due to effective deception measures, Army Group Centre contained just *two* understrength panzer divisions. By then the Germans had about 30 panzer divisions. 20 were on the Eastern Front. 10 were in the west.[12] The Red Army had about 30 'tank corps' of equivalent size. At least 20 of them were committed to Operation Bagration.

The Germans did reinforce Army Group Centre with panzer divisions from *within* the Eastern Front. However, Bagration opened 16 days after D-Day. The Wehrmacht (strictly, the Wehrmacht and the Waffen SS) could not move panzer divisions *to* the Eastern Front. Another ten panzer divisions in Army Group Centre, before Bagration opened, would have made a huge difference to the outcome.

The Second World War was not decided entirely by armoured divisions. However, the Wehrmacht was strategically overstretched from D-Day onwards. One could say that that overstretch became a reality somewhere in Russia in late July 1944. We shall see another interpretation shortly.

Operation Bagration was massively successful. It drove the Red Army west to the Baltic, Poland and Romania. However it does *not* demonstrate a Soviet mastery of the operational level of war. From the summer of 1943 the Red Army outnumbered the Wehrmacht by almost two to one. For a major operation it could easily muster a million *more* men than its opponents. The Red Army had some useful operational theory. However, as we have

12 An 11th (by chance the 11th Panzer Division), operated in Army Group G in southern and south-eastern France throughout the campaign. It is not included here, nor later.

seen, it could get things badly wrong. It is, arguably, easy to succeed at the operational level if an army has three things. One is having (say) a million men more than the opposition. Another is having several more armoured formations. The third is having competent commanders and adequate doctrine. By the summer of 1944 the Red Army had all those things.

We now turn to Normandy. In 2010 the Swedish writer Niklas Zetterling published an immensely detailed analysis of the German forces actually deployed in Normandy, and the losses they suffered. His book was republished in English in 2019. Among many other sources, Zetterling consulted over 100,000 pages of German wartime archives.[13] In some instances he accounted for every last tank within a division. His book exposes several long- and strongly-held myths and errors about the Normandy campaign. Readers who may doubt what follows are strongly advised to read Zetterling's work.

As early as *February* 1944 Montgomery intended that the British Second Army, in the east, would protect the First US Army as it cleared the Cherbourg peninsula and then broke out.[14] It did just that. At one stage in July all ten panzer divisions were engaged against the British and Canadians or (in one case) still moving up to the front. (9th Panzer Division arrived between about 28 July and 6August).[15] As the Americans broke out south of St Lô in late July, Hitler ordered all available panzer divisions to attack westwards, in daylight. Dogged American defence, massive artillery support and significant air attacks halted them.

The defeat of that Germain counterattack, at Mortain from 6 to 13 August, is perhaps the moment when strategic overstretch bit hard. On 1 August Patton had stood at the bridge at Avranches, within sight of the Atlantic. He had just taken command of the Third Army in Normandy. He rapidly broke out and reached the River Seine below Paris on 18 August. The last available panzer divisions, both Wehrmacht and Waffen SS, had been committed at Mortain on 6 August. There had been nothing to stop Patton.

Montgomery and his subordinate General Omar Bradley (who at the time commanded the 12th US Army Group, of First and Third Armies) had botched the encirclement which we know as the Falaise Pocket.

13 Zetterling, *Normandy 1944*, 7.
14 Brooks, *Montgomery and the Battle of Normandy*, 20.
15 Zetterling, *Normandy 1944*, 292-3.

Montgomery did not actually tell Bradley that he intended to encircle the relevant forces and destroy them. He did not actually order Bradley to link up with the Canadians and Poles advancing south from Falaise.[16] Bradley, however, must take some of the blame for the fact that the pocket was not closed effectively. Patton was already many miles away to the east, with two corps. Not more than 640,000 Germans served in Normandy. Of those who fought in divisions (about 77 per cent), 206,000 were killed, wounded or became prisoners of war. About 251,000 escaped.[17] Conversely of 2,336 tanks and self-propelled guns sent to Normandy, 1,500 or so were lost. Only about 230 remained with front-line units but, importantly, another 600 or so were in repair units. They would return to the front line in due course.[18] Similarly only two of 20 formation HQs were lost.[19] The remaining commands could be rebuilt.

The objective of Operation Overlord was to gain a lodgement on the continent of Europe. It had been expected to take 90 days. Paris fell on 24 August, the 80th day. The lodgement had been gained. The Allies took about 143,000 prisoners, a great many of them in the Falaise pocket. Furthermore, the Wehrmacht was defeated so heavily that it rapidily evacuated almost all of France and Belgium. If that was not a great operational success for the Allies, what is?

It is clear that at some stage the German high command planned and ordered that evacuation of France and Belgium. No mention of any relevant order is to be found in English-language histories. Why is that?

There was then considerable discussion and disagreement between Montgomery and Eisenhower as to how to complete the campaign in northwest Europe. At one stage Eisenhower characterised the rest of the campaign as the advance to the Siegfried Line, breaching it (and crossing the Rhine), and then exploiting. That is not an operational plan. It is simply analysis of the mission. Montgomery favoured a narrow thrust. Most of the British and American forces would advance north of the Ardennes and hence the Ruhr. Eisenhower's eventual plan saw the British and Canadians,

16 Ibid, 280-2, 286-90, 301, 306.

17 The remaining 35,000 or so were not involved in the Falaise pocket.

18 Zetterling, *Normandy 1944*, passim.

19 Friedrich Hayn, *Normandy. From Cotentin to Falaise, June-July* 1944. Lyndon Lyons trans. (Oxford: Casemate Publishers, 2022). 130. Coincidentally they were the HQs of the LXXXIV (that is, 84th) Army Corps and the 84th Infantry Division.

reinforced by General William Simpson's Ninth US Army, advance north of the Ruhr. The remainder went around the south. The main effort, as we now describe it, would be with Ninth Army as it enveloped the Ruhr. It would link up with the American First Army which encircled the Ruhr from the south.

Operation Market Garden, the Arnhem operation, will be considered in Chapter 8. One small detail merits attention here. During the operation Brigadier General James Gavin, commanding the US 82nd Airborne Division, received an engraved card inviting him to the formal opening of the British Guards Armoured Division officers' mess: presumably the 'Eye Club' in Brussels. Gavin refused to attend whilst his soldiers were in combat. At that point the Guards Armoured Division was fighting alongside the 82nd Airborne Division in the Nijmegen area. It is an interesting insight into respective military cultures.

The ten panzer divisions which escaped Normandy were refitted. In December eight of them attacked through the Ardennes, in what became the Battle of the Bulge. Patton turned the Third Army north through ninety degrees and counterattacked incredibly quickly. It was a triumph for his Army's intelligence staff, foresight, and anticipatory planning. The German panzer forces were badly mauled and withdrew. It was a major battle: considerably more than 10,000 Americans died. Very few British or Canadians were involved.

Six of the ten panzer divisions were then sent east. The Canadians and British breached the Siegfried Line and crossed the Rhine in Operations Veritable and Blockbuster, then Operation Plunder. Veritable, in February 1944, was a slow, grinding struggle; not least because the Germans committed three of their four remaining panzer divisions. The Ninth US Army had been due to attack at the same time. The attack was delayed by major German flooding operations. That allowed the Germans to commit most of their armoured reserves against the Canadians and British.

When Simpson did attack, 15 days later, he faced little opposition. The American First, Third and Seventh Armies crossed the Rhine further south. The Germans had run out of armoured divisions. Montgomery had five armoured divisions. The Americans had 17. The rest of the campaign in the west was a six-week operational exploitation.

Operation Bagration opened 16 days after D-Day. The Soviet Vistla-Oder Operation of January 1945 opened as the Battle of the Bulge wound down. Stalin and Zhukov had not been confident that the Red Army could succeed, given the forces available to the Germans.[20] In both cases the Soviets attacked once they had seen that the western Allies were tieing down significant armoured forces away from the Eastern Front.

After crossing the Rhine, the British advanced to the Baltic. The Canadians cleared Holland. The Americans advanced east to overrun the rest of Germany: Patton reached Czechoslovakia. It worked. It was not elegant nor clever. Allied operations after Normandy reinforce the idea that the campaign or theatre level was not well understood in Western armies.

Irregular warfare played a major role in the campaigns of the Second World War. In the Soviet case the irregular forces were largely partisans operating behind German lines. They were organised and directed from Moscow. Many of the partisans were soldiers who had escaped capture in the great encirclement battles of 1941 and 1942. In other countries resistance movements were organised by the American Office of Strategic Services (OSS) or the British Special Operations Executive (SOE). At first, the main activities of such resistance movements were espionage and some sabotage. The destruction of the Vemork heavy water plant by Norwegian commandos in 1943 had a significant effect on German atomic weapons development.

When the Allies advanced towards a given country or region, the emphasis switched to guerilla warfare behind enemy lines. Where resistance movements rose up too early the consequences could be swift, brutal and tragic. The fate of French resistance forces in the Vercors in July and August 1944 is an example.

On the night before D-Day, French resistance groups cut railway lines in 486 places in northern France. On 7 June 26 rail routes were still blocked. That significantly disrupted the movement of armoured forces to Normandy. Some panzer divisions reported being attacked from the air during their road moves. That overlooks the fact that they were only moving by road because the railway lines had been cut: *largely by the Resistance.* In total railway lines were cut 577 times. 1,500 locomotives were damaged. The damage affected three quarters of the trains available in northern France. History books generally don't mention that.

20 Willmott, *The Struggle for Europe*, 704.

The Wehrmacht considered that wheeled formations could cover 300 kilometres per day on the march.[21] Tracked formations could cover 200. The 2nd Panzer Division was ordered to Normandy on 9 June. Its wheeled elements moved 400 kilometres, from Amiens via Paris, in two nights.[22] Much has been made of the fact that the threat of air attack forced German armoured divisions to move to Normandy by night, which delayed them; but in practice 2nd Panzer Division could not have moved any faster in daylight. When examined in detail, most of the panzer divisions were not delayed significantly. They took few losses in the process.[23]

Hence, for two reasons, the impact of air attack on the movement of armoured formations to Normandy is generally exaggerated. One reason is that much of the damage to the railways was due to sabotage. The other is that moving by night was just as fast as moving by day. That, however, is not the whole story. Amiens is about 250 kilometres from Normandy. 2nd Panzer Division had to travel 150 kilometres further, via Paris, because the main bridges over the lower Seine had been destroyed by air attack. Its tracked units moved by train, and some of them were significantly delayed. Some aspects of air attack did cause disruption.

Resistance forces naturally thought a lot about what political conditions they wanted after the war. In France the dominant group were right-wing, nationalist, chauvinist and often misogynist (despite the role of female *résistantes*). Not surprisingly, Gaullism dominated post-war France. In Yugoslavia the SOE eventually supported the Tito's Communists, rather than the pro-royalist Chetniks. Communist partisans were particularly active in both Malaya and Vietnam. In all these cases resistance groups had a significant impact on the post-war political environment.

We can make several observations about the operational level the Second World War. Firstly, the war shows a growing awareness of the campaign and theatre level in several countries. Armed forces generally became better at planning and conducting campaigns.

At sea, the main strategic output of all campaigns was, in practice, to launch and support operations on land; or occasionally in the air. That may sound belittling, but appears to be true. The two largest naval campaigns,

---

21 Eike Middeldorf,. *Handbuch der Taktik. Für Führer und Unterführer*. (Berlin: E. S Mittler & Sohn, 1957). Author's translation, 103.

22 Zetterling, *Normandy 1944*, 275-6.

23 Zetterling, *Normandy 1944*, 98.

the Battle of the Atlantic and the campaign in the central Pacific, brought land forces to the point from which the enemy homeland could be assaulted.

Time and time again, air superiority was critical to naval success. By the end of the war, aircraft carriers had supplanted battleships as the major class of warship. The operational effect, however, was maritime; not aerial. Those ships had to reach their objectives. Aircraft could not replace them.

The effect of long-range bombing has been grossly overrepresented. There was little direct effect at the strategic level. Two atomic bombs were instrumental in Japan's surrender. That saved the lives of perhaps a million Allied servicemen. Planning for the invasion of Japan was already well underway. The First Canadian Army was being shipped back across the Atlantic to take part. Five RN aircraft carriers were already operating in the Pacific. The question was not *if* Japan would be overrun; but when and by whom.

The atomic bombs *may* have resolved, or settled, Japan's participation in the war. But more than anything else they represented overwhelming, strategic, one-sided, technological advantage. They were dropped by Army aircraft. Those aircraft might conceivably have belonged to the Navy. To use the bombing to justify an independent US Air Force was a triumph of domestic service politics. Calling long range bombing 'strategic' is an important yet misleading part of air force propaganda. Damage caused by munitions is tactical, even if occurs in the enemy's capital.

Apart from the atomic bombs dropped on Japan, air forces had little or no *direct* strategic effect. They did not wreck the German war economy. Conversely the *disruption* which long range bombing caused to Axis war economies, and armed forces, was massive. In Europe probably the most important effect was to gain air superiority in the west. In doing so, Allied air forces made a major contribution to Allied operational (and hence strategic) success. Air superiority at the theatre level was critical to success. Support to surface forces, both naval and military, was highly important. At times it was critical. As we shall see, Slessor was largely right. So were the older generation of Luftwaffe pilots. Trenchard's staff, Douhet, Mitchell, Wever and Harris were not.

On land, the decisive war-winning weapon was the armoured division. It could not work well without (at least) air parity. It had to be a well-trained, well-led, all-arms team. It worked best in deep, operational-level encirclements or envelopments. Given those circumstances, armoured forces were repeatedly decisive: operationally and hence strategically.

Compare, for example, the events of the campaigns in western Europe in 1940 and 1944-5 with those of the 100 Days in 1918.

There is a tendency to criticise the Wehrmacht for things which do not win campaigns and battles, such as intelligence and logistics. The Wehrmacht seems to have focussed fairly ruthlessly on things which *did* win campaigns and battles. Anything else was, by default, secondary. Campaigns and battles are won by fighting and manoeuvre. Intelligence and logistics are enablers. So we should ask whether modern, western armies are actually properly focussed on what leads to operational and tactical success.

However: if you do not destroy them, you will have to fight them again. To use just the examples in this chapter: Montgomery could not destroy the Afrika Korps at El Alamein. He had to fight it again in Tunisia. Clark chose not to destroy the German Tenth Army south of Rome. The Allies were still fighting it 11 months later. Montgomery and Bradley did not destroy the German armoured forces in the Falaise Pocket. They had to fight them again in the Battle of the Bulge. Major encirclements by armoured formations can have considerable operational effect. They will probably not be decisive of themselves unless they destroy the forces they encircle. To repeat: if you do not destroy them, you will have to fight them again.

The Soviets were not the masters of the operational level of war. Any army *should* look good when it can outnumber its opponent by a million men; unless it is grossly incompetent. The Soviets achieved some level of competence by the end of the war.

Finally, we have remarked that historians have missed something significant five times in this chapter.

8

The Second World War: Tactics

Chapter Eight considers the tactics of the Second World War. We cannot consider every army, navy or air force. Nor can we discuss every battle or engagement. What follows is a necessarily selective look at warfare at low level between 1939 and 1945, starting in the air.

On 10 May 1940 the Luftwaffe lost 308 aircraft over France and the Low Countries. It was the Luftwaffe's biggest ever loss on a single day. Those losses were far bigger than on any day in the Battle of Britain. Yet they gained air superiority for the whole campaign. The Luftwaffe concentrated force ruthlessly over its army's main effort, and seemed to have undertaken very little long-range, 'strategic' bombing.

The BEF and its air component were summarily kicked out of mainland Europe in June. The Army quickly ordered an investigation, conducted by what became the Bartholomew Committee. It reported on 2 July. Regarding air support, the report said 'that urgent action is required to place co-operation between the two services on a better basis.'[1] It highlighted that the Luftwaffe liaised far more effectively with the Wehrmacht than the RAF had with the British Army.

The RAF quickly rejected those findings. As we have seen, they were already bombing Germany. For the next few years air support to the Army (for example, in the Western Desert) was poorly coordinated. Nevertheless, by D-Day the RAF (and the USAAF) had created organisations that could do largely what the Army had asked it to do in 1940.

The Fairey Battle light bomber demonstrates much of what the RAF had got wrong in the 1930s. In 1940 level bombing with single-engined light bombers, such as Battles, simply did not work. Battles were shot down

---

1 Final report of the Bartholomew Committee on lessons to be learnt from the operations in Flanders.pdf (archive.org) accessed 1401 hrs 15 February 2024.

in very large numbers. They were slow and vulnerable. They have been described as 'obsolete in 1939'. That is a strange description for an aircraft which had only entered service in 1936. The RAF had clearly made a wrong choice. More broadly, the RAF did not really want to provide offensive air support to the Army at all.

The Battle was the first aircraft to use the Rolls Royce Merlin engine. It was rapidly standardised. The Merlin was used on Hurricanes, Spitfires, Lancasters, Mosquitoes and several other types. Packard built Merlins under license for the American P51 Mustang. The Mosquito was probably the most successful British aircraft of the War. By 1944 the RAF could destroy individual factories with absolute precision, using a few Mosquitoes and 10-15 tons of bombs. However the RAF typically continued to use hundreds of Lancasters, and hundreds of tons of bombs, to plaster cities by night and miss important targets. The RAF had had no stated requirement for the Mosquito.

On the night of 16-17 May 1943, 19 Lancasters of 617 Squadron RAF attacked four dams in Germany using special 'dam busting' bouncing bombs. Two of the four dams were destroyed. Eight aircraft (42 per cent) were lost, and *the operation was never repeated*. Yet, weirdly, the Dams Raid is still used today to commemorate the supposed nobility of long-range bombing. This is an example of organisational hypocrisy which is obvious, yet totally overlooked. The undoubted heroism of the crews should not be used to cloud objective assessment.

In June 1942 a small force of USAAF Liberators attacked the Ploieşti oilfields from north Africa. The target was poorly defended, but damage was light. The next attack was 14 months later. In the interim the Luftwaffe and the Romanian air force had reinforced the air defences with three fighter groups, several hundred antiaircraft guns, and radar. 162 Liberators attacked the target at low level on 1 August 1943. Surprise was lost due to poor navigation and COMSEC. Only 88 aircraft returned to Libya, of which 55 were damaged. Surprise is a principle of war. In this case surprise was lost both operationally (by attacking in 1942) and tactically (through poor COMSEC and so on). With surprise lost, attacking unescorted at low level and in daylight resulted in horrendous losses. Damage to the oilfields was negligible.

Bombing is a raiding tactic. Surprise is a key feature in the success of raids. Air forces are not excused from adhering to the principles of war. However, much more positively, in 1944 Montgomery wrote:

- That the enemy's air force must be subdued before a land offensive is launched.
- That a retreating enemy offers the most favourable targets to air attack. In such conditions, action by the air forces can turn retreat into rout. It may be decisive.
- Critically, that the moral effect of air action is very great and is out of proportion to the material damage inflicted.[2]

It seems that Slessor and Montgomery would have agreed.

We now turn to naval warfare, and start with naval aviation. In May 1941 *Bismarck* was torpedoed by carrier-based aircraft which had located the target using airborne radar. *Bismarck's* steering was damaged. That led to her being destroyed by gunfire the following day. Airborne torpedo attacks were not easy. In December 1941, off Malaya, HMS *Repulse* dodged 19 Japanese torpedoes before being hit. In the same attack HMS *Prince of Wales* was hit by just one of nine torpedoes launched at her. It made her lose power and list to port. Her starboard antiaircraft battery could no longer depress to sea level. That left her vulnerable to further attacks from the starboard, which sank her.

In February 1942 *Scharnhorst, Gneisenau, Prinz Eugen* and escorts made a dash from Brest in Brittany through the English Channel. They were heading for the German North Sea ports. German EW units jammed British radar and radio frequencies in a surprise electronic attack. The few RAF anti-shipping aircraft had recently been sent to North Africa. RAF maritime strike wings were then assembled, several months after the German warships reached Germany. Maritime attacks on German coastal shipping then became very effective.

On 4 April 1944, 40 carrier-based Royal Navy dive-bombers attacked the *Tirpitz* in Altafjord in northern Norway. Surprise was complete. Watertight compartments were not closed. Antiaircraft guns were not manned. *Tirpitz* was hit 15 times. She was heavily damaged, but did not

---

2 Field Marshal B L Montgomery, *Some Notes on the Use of Air Power in Support of Land Operations.* (HQ 21st Army Group, Holland, December 1944), 8.

sink. Two RN fleet and four escort carriers had launched the sorties. The Navy attacked the *Tirpitz* several more times that summer, and scored some hits. However the bombs were not heavy enough sink such a well-protected battleship. The RAF sank *Tirpitz*, on the third attempt, with 32 Lancasters using 12,000 lb 'Tallboy' bombs on 12 November.

Navies had been well aware of the threat from aircraft. The Royal Navy had started to order dedicated antiaircraft cruisers in 1937, largely for convoy and carrier escort. The antiaircraft armament of virtually all classes of warships increased considerably through the war. Escort carriers were built in large numbers, largely to provide fleet and convoy air defence.

Airborne torpedoes had a range of perhaps 1,500 yards. Surface- or submarine-launched torpedoes had a range of about 4,000 yards, with a run time of about four minutes. Japanese 'Long Lance' torpedoes could reach 44,000 yards. However, with a run time of maybe 30 minutes to that range, the target had to be careless or unlucky. In the Great War destroyers typically had two or three torpedo tubes. By the end of the Second World War they had eight or ten. By 1944 British destroyers were jamming the guidance signals of German anti-ship guided bombs. They were usually, but not always, successful.

Naval gunfire became deadly once it was linked to radar fire control systems. In 1942 Japanese battleships could hit another ship with perhaps the fourth salvo. Using radar, US battleships could hit with the first salvo. On 15 November 1942 the US battleship *Washington* engaged and sank the Japanese battleship *Kirishima* off Guadalcanal, at night, at ranges from 8,400 down to 4,800 yards, firing 62 rounds within seven minutes (and despite having to cease fire for 90 seconds).

We now look at army tactics. In an interesting insight, in 1940 the British observed that German attacks in Belgium and France always failed where the defender's Bren (light) machineguns were sited to fire in enfilade. That is, to protect the adjoining section or platoon. This says something about the physical cohesion of the defence, which we shall return to later.

Montgomery's attack at El Alamein opened on the night of 23-24 October 1942. Four British divisions attacked on a frontage of 12,000 yards, widening to 18,000. The British had just over 1,000 tanks; the Germans and Italians about 540. It was planned to take 12 days. It took 12 days.

456 field guns, but just 48 mediums, fired on the attacking sector. All of the mediums were used in one group, which neutralised one enemy battery

at a time. The field guns fired a preliminary 15-minute counterbattery fireplan, in which the 72 guns of each division's artillery concentrated on one enemy battery every three minutes. That equals a concentration of 12 guns to each defending gun. The main fireplan then launched the infantry onto the objective, 6-8,000 yards distant. Return fire was described as 'feeble'. The initial attack was broadly successful. The 'Blue Line' objective, also called 'Oxalic', was reached in most places. The initial attack could be described as 'Amiens, 1918, with sand'.

However most of the British armour was held not just in different divisions, but in a different corps. It was difficult to clear minefield lanes during the night, so very few tanks got through before dawn. German and Italian armoured forces were held in depth behind Oxalic. Hence, from the first day, the British were trying to fight through a salient between enemy infantry positions, and into Axis armoured forces deployed in depth. The British had a total of 860 guns, but could rarely bring them all to bear into the salient.

For each of the next eight days or so the pattern was broadly the same. A few enemy machineguns, mortars or antitank guns which survived the British fireplan would hold up the advance. It was perhaps not until 1 November that Montgomery refined his methods. That night, for Operation Supercharge, 192 guns fired a barrage and 168 fired in counterbattery. Two New Zealand infantry brigades cleared a gap about 6,000 yards deep. However 9th Armoured Brigade, attacking at dawn, was then held up on an antitank screen. A similar attack on the morning of 4 November broke through. However, that was largely because Rommel had started to withdraw.

The basic problem was that of neutralising individual crew-served weapons. Small arms fire would suppress, but not neutralise. To neutralise a machinegun or an antitank gun, a tank needed to be able to fire HE of about three inch (75mm) calibre or greater. However British tanks could only fire solid shot. Alternatively, infantry units could neutralise machineguns and antitank guns with mortar fire; but British infantry battalions initially only had two three-inch mortars each, and the mortar platoon had no radios.

Once the initial fireplan had been fired, British artillery was not flexible enough to neutralise the remaining pockets of resistance. It should have been easy to neutralise antitank guns, and even the feared 88mm antiaircraft guns, with artillery or (better) mortars. That, however, rarely happened. American M4 Sherman tanks, with 75mm guns which fired HE, had just arrived in theatre. Rommel lost all of his 88s at El Alamein. That was highly significant, but largely unnoticed by historians.

British artillery tactics were evolving. Barrages were generally being replaced by concentrations on successive lines. A barrage is, strictly, uneconomical: many of the rounds are fired at areas where there are no enemy. However, as we have seen, successive concentrations did not in practice neutralise all the enemy's support weapons. The 2nd New Zealand division used barrages throughout the war. Its attacks 'seldom failed.'[3]

British forces could clearly break into a defensive position and fight their way through. What they did not do was break right through; that is, break out. British forces would typically do what they were ordered, if they could. They then expected to be told what to do next. That applied right up to corps level. British tactics remained pedestrian, throughout the Army. Command was over-centralised. It worked, but not well.

In Burma the Japanese held an initial advantage in jungle fighting. They were tough, highly disciplined, and could survive with a minimum of supplies. The Japanese were obedient, unimaginative and inflexible. They excelled at meeting engagements, which developed into envelopments. By doing so they cut the trails behind their enemies, causing them to panic and fall back.

Initially both the Japanese and the British tended to operate with their forces widely dispersed. Thus at higher tactical levels their defences lacked mutual support: that is, cohesion. That was an advantage to the Japanese whilst they held the initiative. When they lost it, their formations and units could be isolated and defeated in detail. That is broadly what happened in the Kohima-Imphal campaign.

The British did not really consider jungle fighting until forced to, in late 1941 and 1942. It took them until 1944 to adapt to jungle conditions and gain the confidence to take on the Japanese. British tactics in the first Arakan campaign of 1942-3 were often unimaginative; frontal; and often failed.

By early 1944 the British had the tactical and then operational initiative. Japanese tactics were frequently predictable. Artillery was often ineffective when firing through the jungle canopy. The situation required the coordinated use of small arms, sometimes tanks, and well-executed

---

3 Pemberton, *The Development of Artillery Tactics and Equipment*, 279.

minor unit tactics. British, Indian and African units learned how to beat the Japanese through hard experience.

Much the same applied to American forces in the Pacific. There was little sharing of tactical lessons between the Allies. Americans learnt to avoid frontal attacks and exploit flanks. In defence they learnt to encourage the Japanese to attack frontally, and then destroy them with massed firepower.

It is difficult to appreciate just how bad Red Army tactics were. Russia had entered the Great War with nineteenth-century tactical methods. It had never learnt better. It fought the Second World War with the same methods but 1940s equipment. It lost commanders so quickly that there was virtually no-one left to teach it how to fight better.

As an illustration, near Gusevo in November 1942 (during Operation Mars), the German 110th Infantry Division had to defend a 30km sector. That meant just a single platoon in reserve per battalion; a company per regiment; and possibly one battalion for the division. Every front-line company (21 out of 27) would have a sector of almost 1,500m. The Soviets failed to break through. A few days later four Soviet rifle divisions and two tank corps attacked near Osuga on a 4km sector. They were supported by 2,500 guns, mortars and rocket launchers. They were faced by elements of two German regiments. The Soviets attacked for three days and gained less than 1,000m. They didn't capture a single village. Casualties, of both tanks and infantry, were over half of the whole force. German casualties in Operation Mars were typically one or two hundred per division per week: not particularly high. Compare that with German losses on the Somme.

At or about the same time, General Gotthard Heinrici was commanding the German Fourth Army. He was defending a 200km front astride the Minsk-Moscow highway with ten divisions. The Red Army attacked five times on the 20km sector nearest the road. On each occasion they attacked for five or six days at a time, and typically at three predictable times per day. The Red Army deployed 20 to 38 divisions in each attack. German casualties were typically equivalent to about one battalion per day of battle. The Red Army did not get through. Soviet losses were often 12 times, and sometimes 18 times, higher than the Germans'.

At Kursk the Germans were outnumbered by about 2.45:1 in men and 3.12:1 in tanks. They attacked. They inflicted about 4.25 times more casualties on the Russians than they suffered, and knocked out 7.07 times as many tanks.

Overall, the Wehrmacht considered that conducting a successful defence did not depend particularly on the strength of the defending force. Nor did it depend on the force ratio. It depended, in the first instance, on the *ratio of force to space*. The defenders had to have enough forces to form a coherent defence across the whole sector, with some reserves. Given that, they could defend successfully against greatly superior enemy forces. The relevance of the earlier remarks about British Bren guns in 1940 is now apparent.

German assault gun battalions would typically destroy about 10-15 Soviet tanks for each assault gun lost. In poor battalions it would be six or seven to one; in good battalions up to 22 to one. Individual assault gun commanders knew that they could safely attack if four enemy tanks were visible. Five would be tricky, but possible. Importantly, all German assault guns could also fire HE.

At low level, Red Army units could be dogged in defence. Their artillery fire was generally heavy but inaccurate and ineffective. It was often predicted, not observed, and based on poor intelligence. Soviet units' tactical reporting was often non-existent or poor, resulting in bad decision-making. COMSEC was frequently appalling. Yes: the Soviets sometimes managed to deceive the Germans. Their officers were often poorly trained. Attempts to motivate, normally based on Soviet ideology, were often ineffective. Draconian punishments did not prevent ill-discipline, but resulted in poor morale.

Clearly those are generalisations. The examples above tend to emphasize the best and overlook the worst German performances. There was some Soviet improvement later in the war, and the overall quality of German forces declined. Yet even at the Seelow Heights east of Berlin in April 1945, and then within the city itself, German units often repeatedly outperformed their opponents by a large margin. In the end, however, the Red Army's quantity had a quality all of its own.

Anyone who thinks that a T34 was a good tank has not seen a T34. It had a reasonably powerful, reliable engine and a fairly good gun. The early models also only had a two-man turret, no turret basket and no turret crew seats. Initially few had radios. The optics and fire control system were appalling. When it entered service, its main advantages were that it weighed 26 tons and had sloped armour. It was therefore well-protected. The best German tank, the Panzer IV, weighed just 21 tons.

The T34 was in service when Germany invaded Russia in 1941. The German Army produced the Panzer V (Panther) in time for Kursk two years later, specifically to counter the T34. It was an extremely good tank once production issues (such as oil seals and gaskets) were overcome. That took a few months. The Panther was considerably better than early T34s and was never upgraded. However it was, if anything, too heavy. It entered service at 44 tons.

British tank design and production was woeful for several years. Over 3,100 tanks (including Covenanters, Cavaliers and Centaurs) were built, found to be inadequate, and scrapped before entering service. That changed after late 1941 when W A Robotham, a Rolls Royce sports car designer with no experience of tanks, was put in charge of tank design. Robotham had proposed the use of Merlin engines for tanks. They were very powerful, yet compact. Robotham redesigned the Merlin as the Meteor and used it in three tank projects. They became the (nominally) 30-ton Cromwell, the 40-ton Comet and the 50-ton Centurion. They appeared in 1943, 1944 and 1945 respectively. The Centurion entered service a few days before the end of the war. Its last battle was in 1991.

The British 2 pdr antitank was very effective in 1940. In the Western Desert it was, critically, outranged. It was also not powerful enough to defeat the armour of newer German tanks. The 6 pdr arrived in time for Alamein. The 17 pdr appeared a few months later, in Tunisia. The 17 pdr was an outstanding gun which could defeat the frontal armour of a Tiger at 1,500 metres.

However, we should not just consider the design of the tank and the gun. We also need to look at numbers and organisations. The US Army standardised on Shermans until they could begin to replace them with M26 Pershings. The German Army wanted as many Panthers as it could get. It never got enough. By mid-1944 half the fleet (one battalion of the two in most panzer divisions) still had Panzer IVs. There were only ever 15 battalions of Panzer VI 'Tigers'. In practice that meant perhaps half a dozen in the east, typically two or three in western Europe, and perhaps one in Italy at any one time. Germany built 5,976 Panthers but only 1,843 Tigers.

At the end of the war there were more assault guns in German service than any type of tank. Similarly there were 52 tank battalions in US armoured divisions, but 65 for use with infantry divisions. The British 21st

Army Group had five armoured brigades in armoured divisions, but eight armoured or tank brigades for use with infantry divisions.

Terrain can be a problem if forces are not trained for it. That was especially true in southern Italy. In many areas the mountains dropped straight to the sea. The rivers were often narrow, but deep and very fast. As always, German defensive tactics were based on a covering force, the main line of resistance, and counterattacks. The skill lies in the relation between the three elements, and between them and the terrain. In December 1943 the Bernhardt Line marked the covering force area for the main Gustav Line at Cassino. The US 36th Infantry Division committed 12 American battalions and an Italian regiment, and fought for seven days, to drive back three German battalions in the covering force area.

Several British and Dominion commentators remarked that American minor unit tactics were poor. More generally one (British) Indian Army general described the first *three* battles of Cassino (by American, British and New Zealand forces) as 'military sins'. His experience on the Indian North-West Frontier probably informed his opinion. The Polish Corps then studied the terrain at Cassino in great detail. Admittedly they learnt from the experience of the three previous failed attacks. They concluded that the key terrain was not the Abbey, nor Castle Hill below it, nor the town below that (see Figure 8-1). It was the narrow ridge behind the Abbey. Seizing the ridge cut off the Abbey. Taking the abbey made the town untenable. Seizing the ridge also opened the Liri valley and hence the road to Rome.

The German infantry positions on the ridge were the main line of resistance for the Gustav Line. Section outposts were sited on the forward slope, and re-occupied whenever possible. The ridge is steep but narrow. Reserves on the reverse slope were effectively immune to indirect fire, but close enough to counterattack quickly. It was a hugely strong position.[4]

The Germans used assault guns. The Americans, Canadians, British, New Zealanders, Poles and French used Shermans. All had 75mm guns which could fire HE. However, the terrain made it very difficult to use more than a few tanks at any one time. Due to the knife-edged ridges behind the Abbey, it was technically very difficult to use artillery effectively. The Poles therefore used most of their antiaircraft gunners to man mortars.

4 I have visited the Cassino battlefields five times.

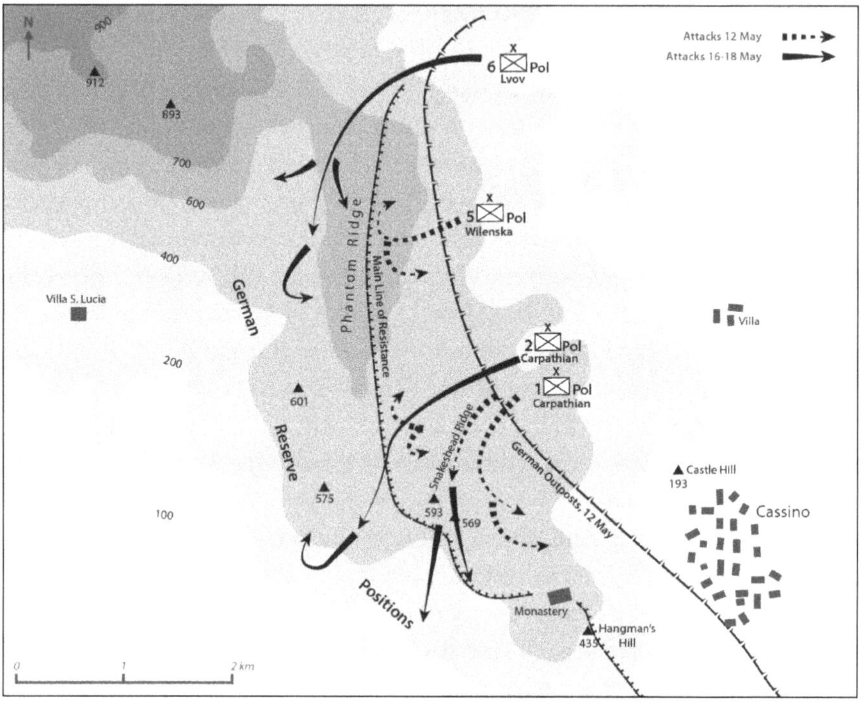

Figure 8-1. Cassino and Cassino Abbey, May 1944

In January 1944 a New Zealand brigadier remarked that American front-line GIs at Cassino were in bad shape. (Several had to be carried out of their positions by the Indian soldiers who relieved them.) However, he said, American senior officers didn't know that because they didn't go forward often enough. We shall return to this subject later.

After Cassino and Anzio, western attention shifted to Normandy. Montgomery's hand-written one-page concept for the landing has its last word in capitals, underlined three times: SIMPLICITY. Yet the landing plan for each of the seven American assault battalions on Omaha beach was different. One very experienced amphibious commander has described the plan as 'the work of a drunken spider' and referred to 'the seductive appeal of complexity'. Boat handling and seamanship were poor. The landing forces became hopelessly intermingled. Amphibious ('DD') tanks were poorly handled. There were no armoured engineer tanks. Therefore obstacles had to be cleared by hand, on foot, under fire. Of the seven or eight reasons for the high casualties on Omaha, perhaps two were beyond the US Army's control.

Casualties were relatively light elsewhere but deserve analysis. The British lost 379 tanks on D-Day. 452 crewmen were killed: roughly the same as all infantry killed on the British and Canadian beaches. Comparison with Omaha beach suggests that a few hundred armoured casualties saved thousands of infantry casualties.[5] In particular, six battalions of specialist armoured engineer vehicles of the 79th Armoured Engineer Division (79 AED) landed with the first waves of infantry. Crew losses to 79 AED were surprisingly light.[6,]

The early fighting north of Caen and in the hedgerows further west was bitter. The hedgerows were described as 'the graduate school of the US Infantry'. German machineguns were highly effective. Their mortars were a constant hazard. About 70 per cent of British infantry casualties were caused by mortar fire. That amounted to perhaps 25,000 casualties in the first seven weeks of the campaign. Much the same probably happened to the Americans.

Almost every attack, British and American, needed tanks. Artillery alone rarely got the infantry onto the objective. British doctrine had stressed that 'it is the platoon commander's duty to lead the assault.' That had to be modified. The infantry could simply not afford the losses. 40 years later, infantry subalterns were told that their place in the assault was 'between and slightly behind the assault sections'. It didn't stop them getting Military Crosses. It did stop them being killed or injured in such large numbers.

Writing in his diary on 19 July, Montgomery wrote that '[t]he bigger American casualties are due to their lack of skill in fighting.'[7] That reflects a common British and Dominion view. We shall return to this later.

On 13 June the veteran British 7th Armoured Division had been mauled by a few Tigers of 101st SS Heavy Panzer Battalion at Villers Bocage. The 17 pdr antitank guns of the divisional antitank battalion were quite capable of destroying Tigers. Infantry battalion 6 pdrs could also do so if used properly. Villers Bocage is a clear example of poor unit and formation

5 This is a correction. Regrettably the equivalent passage in my earlier book, *Hall of Mirrors* (page 151) was incorrect.

6 Eight officers and 23 soldiers were killed on D-Day. *The Story of 79 Armoured Division, October 1942 – June 1945*. Author(s) not named. Printed in Hamburg, July 1945. 53.

7 Brooks, *Montgomery and the Battle of Normandy*, 223.

tactics. The divisional reconnaissance battalion failed to give warning. The antitank battalion failed to protect the armoured units. The leading tank and infantry units were caught halted in column of route. Caution characterised British handling of armour for a long time afterwards. Villers Bocage became an SS propaganda victory. The Anglo-Saxon description should be rather more vulgar.

The Goodwood and Cobra battles deserve comparison. By 29 June 1944 six, and elements of a seventh, panzer divisions were facing the British (the rest of the seventh was facing the Americans).[8] By late July there were ten, of which two were facing the Americans. It was probably the highest concentration of panzer divisions on one sector, ever. Heavy bombers were used at both Goodwood and Cobra.

At Goodwood, southeast of Caen, a railway line marked the end of the bomb line and, effectively, the limit of field artillery range.[9] It is generally forgotten that the attacking divisions completely overran the front-line (Luftwaffe Field) division: Amiens, 1918 yet again. Much of 21st Panzer Division, and elements of two more, were deployed behind it. In broad terms the British armoured divisions did not progress beyond the railway line on the first day. Goodwood resulted in an advance of about 7,000 yards that day, and 10,000 by the third. 7,000 yards in one day is impressive by the standards of the Great War. However, comparing the effective range of the 25 pdr with the Great War 18 pdr tells us that the British had, effectively, made no progress.[10]

Montgomery had no intention of breaking out at Goodwood.[11] He said that there was no particular objective to aim for in that sector.[12] That was consistent with his intentions ever since April. He *did* anticipate that armoured car units would exploit any success. Be that as it may, Goodwood (and its successor, the Canadian Operation Tractable) showed once again

---

8 Ellis, *The Battle of Normandy*, 287. 21st Panzer Division was further east than the area shown on the map.

9 There were actually three railway lines. The one referred to here is the extant Caen-Paris line.

10 Planning ranges were 60 percent of maximum range. For the 18 pdr that was 3,950 yards. For the 25 pdr it was 8,000 yards.

11 *8 Corps Operations East of Caen, 18-21 July 1944. Operation Goodwood. BAOR Battlefield Tour, First Day*. G (Trg) Branch HQ British Army of the Rhine, June 1947. (Uckfield, East Sussex: The Naval and Military Press Ltd, undated), 75 and 84. The reference shows the actual orders given by both VIII Corps and the leading division, 11th Armoured Division.

12 Brooks, *Montgomery and the Battle of Normandy*, 213 and 224.

that British formations could break in to a German defensive position. They could make reasonable progress. But they could, or did, not break through. They never broke out.

The Bartholomew Report had explicitly affected the British Army's approach. After 1940 its doctrine shifted to become more deliberate, cautious and determined.[13] That was fine, as far as it went. Sadly the British probably never learnt how to be more opportunistic. That is, how to exploit. It never learnt how to break out. Not least, by being cautious it *prevented* itself learning how to be more opportunistic.

The American Operation Cobra started on 25 July. The start line was immediately west of Saint-Lô. Like Goodwood, Cobra started with heavy bomber support. It took six days to achieve a breakthrough. The Americans soon had four armoured divisions abreast, attacking through an ever-weakening German defence. By 31 July they had reached the River Sélune, 60 kilometres from the Cobra start line. Then, at Avranches, they broke out.

As mentioned in Chapter 7, the German counterattack at Mortain was seriously disrupted by close air support attacks. The aircraft were principally RAF Typhoons of 83 and 84 Groups, based in Normandy. Their main weapon was the 60lb RP-3 air-to-ground rocket. The British Army Operational Research Group (AORG) visited the scene soon after. Huge numbers of German vehicles had been abandoned. Many were damaged. Many were undamaged, and had fuel in their tanks. However,contrary to pilots' claims, very few tanks had been destroyed.

To quote Zetterling, 'Allied airpower seems to have been misrepresented quite often.' We now know that pilots of close air support missions over-claimed damage to ground targets in Normandy more than five-fold.[14] More generally '... not a single instance supporting the image of extensive losses being inflicted by Allied air forces on German combat units in Normandy has been found.'[15] That includes attack by fighter bombers and so-called 'carpet bombing'.

The effect on the Germans at Mortain was primarily psychological. Many vehicle crews ran away, abandoning their vehicles. That effect,

---

13 Martin Samuels, Clausewitz and the Personality Characteristics of the Battlefield Commander in British and German Military Doctrine, 1918-1941. *War in History,* Volume 30 No 2, 143.

14 Zetterling, *Normandy 1944*, 46.

15 Zetterling, *Normandy 1944*, 41.

however, was real. Those crews *did* run away. They *did* abandon their vehicles. However, in another report the AORG may have stumbled on the root cause. Every single tank that the AORG ever found that had been hit by 60lb rockets was knocked out. The RP-3 seems to have been one of those occasional 'wonder weapons' which were so effective that the enemy quickly learnt not to stay around.

It took American armoured divisions six days to break out in Operation Cobra. However, they did break out. British armoured divisions never did. On several occasions in Normandy, and later, British reconnaissance units found undefended bridges in the enemy's rear. Then their parent formations simply did not exploit the opportunity in time. German formations were adept at establishing ad-hoc defensive screens in depth, and strengthening them before the following morning. Allied forces were not good at realising that there was little, and sometimes nothing, in front of them. The British were the more pedestrian. On the evening of 25 June, during Operation Epsom, the German Army Group B reported a hole in its front line five kilometres wide and two deep.[16] The British simply didn't notice.

Normandy, and the rest of the campaign in north-west Europe, demonstrated that what was needed were well-integrated, all-arms armoured forces. They emerged, as the campaign progressed. In practice the US Army succeeded better than the British. British independent tank and armoured brigades were allocated to infantry divisions as required. One unit (the Sherwood Rangers) fought with 20 different infantry battalions. American tank battalions were generally affiliated to a specific infantry division. Low-level, habitual working relationships were probably generally better.[17]

The Wehrmacht generally avoided tank versus tank battles. They knew, as did the British, that in a straight one-on-one fight the antitank gun generally had the advantage. (The trick in such circumstances, therefore, was to avoid such one-on-one fights.) Tanks were used primarily for attacks or counterattacks. German antitank units were typically very well handled. That was a major factor in the Germans' ability to assemble defensive screens quickly.

---

16 Ellis, *The Battle of Normandy*, 277.

17 I wish to thank to Dr Dermot Rooney for that observation.

British antitank units were quite effective once a deliberate defensive position had been established, but less good before that. Conversely American tank destroyer units were highly variable. Some were outstanding. Many were mediocre or poor. The problem seems to have been their doctrine. Some COs adapted their tactics to exploit terrain and mobility in order to engage the enemy repeatedly from flanking positions. It could be very effective. It was not common.

Allied artillery improved as the campaign progressed. The British had formed Army Groups, Royal Artillery (AGRAs) before Alamein. They were large brigades of medium and heavy artillery which could be switched to support different corps as required. AGRAs provided more flexible control of indirect fire as well as sheer numbers of medium and heavy guns. The record appears to have been a Canadian call for fire on an impromptu (unregistered) target which was answered by 668 guns in 33 minutes. They fired 3,509 rounds, or 92 tons of ammunition, using only five or six rounds per gun. The fire mission lasted for just two minutes. Calls for fire of a hundred guns responding in 12 minutes was fairly typical. By the end of the war counterbattery concentrations of 40 Allied guns to each German gun to one were common.

It could, however, be overdone. The fireplan for Operation Veritable included tank and AA guns, together with huge amounts of field, medium and heavy artillery. AORG later concluded that much of it was unnecessary. At a certain point, all the defenders are suppressed. Most are neutralised. Few are killed. More fire support tends to neutralise the defenders for slightly longer, but kill very few more.

But, critically, if the attacking troops cannot (or do not) keep up with the fireplan, the suppression and neutralisation wear off. That is regardless of the weight of the initial bombardment. In Sicily British gunners had identified that the infantry had to arrive on the objective within two minutes of the fire lifting. If they did not, the bombardment was wasted. The attack would then fail. Divisional attacks could fail even if supported by 15 battalions of artillery. That seems to have happened relatively often in Veritable, and elsewhere. It is strongly reminiscent of the middle stages of the Great War.

Allied armies conducted several major urban operations in northwest Europe. The US Army cleared Cherbourg, Brest and subsequently Aachen. The British seized Le Havre for a cost of 388 casualties. The Germans lost 11,302, mostly as prisoners. The Canadians seized Boulogne, at a cost of 634 casualties. The Germans lost 9,517, again mostly as prisoners. The Canadians then took Calais at a cost of less than 300 casualties. The Germans lost 7,500. American casualties were generally much higher than British and Canadian. That may be partly due to the absence of specialised armour. All those towns and cities were thoroughly prepared for defence.

Two other points should be made. Firstly, notwithstanding that American casualties were generally much higher than British and Canadian, the Germans' (defenders') casualties were always at least 50 per cent higher than the American attackers'.[18] Secondly, American casualties at Brest equate to 3,250 per division. That is very close to the average of 3,291 suffered by German divisions on the Somme.

Operation Market Garden, the Arnhem operation, got under way at 12.35hrs on 17 September. A single armoured division (the Guards Armoured Division) advanced on a one-battalion front, on one road. Later that day an infantry division (the 53rd) attacked on its left flank. The following morning another infantry division (the 3rd) attacked on the right flank. It was scarcely a concentrated effort.

Planners had had seven days to organise the operation. There was not just one road to Nijmegen, and hence Arnhem: there were three. One went through Tilburg. The direct route ran through Eindhoven. A third ran through Deurne. It should have been quite possible to gather three armoured divisions (the British subsequently did) and 79 AED (which they did not.) (It was conducting operations against the Channel ports). The Guards Armoured Division, on the main effort, was apparently lackadaisical.

Is that true? Figure 8-2 shows the advance of XXX Corps. It reached the Rhein at Driel on the evening of D+5. It is clear that some criticism is clearly misplaced. The Guards Armoured Division might, possibly, have advanced faster to Nijmegen. They linked up with the 82nd US Airborne Division there at about 1000hrs on 19 September (D+2). However, neither the Guards nor the 82nd achieved a crossing until 1830hrs the next day. That gave the Germans an extra day. Thereafter the operation got further and further behind schedule. Eventually it was called off. There was no

18 51 per cent higher at Aachen; 73 per cent higher at Cherbourg; 286 per cent higher at Brest and 313 per cent higher at St Malo.

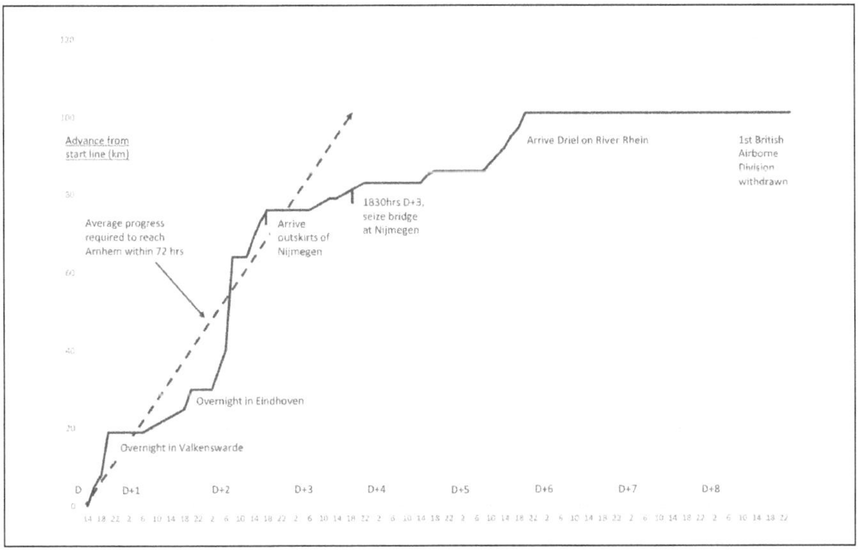

Figure 8-2. Progress of Operation Market Garden, September 1944

advance across the Rhine. Most of the British 1st Airborne Division were killed or captured.

We now consider a number of general issues related to land warfare. For convenience, they are described here in the context of northwest Europe. The British had no flamethrower tanks on D-Day. Ten weeks later they had a brigade of them. Every infantry battalion soon had tracked, carrier-mounted flamethrowers and some manpack equipment. British forces developed highly successful combined tank, infantry and flamethrower attack tactics. Typically, relatively few Germans were killed. Often 30 times as many surrendered. Curiously, British accounts rarely mention the subject.

As the campaign developed, independent American Armored Cavalry Groups (small brigades) were brought into the line. They simply didn't work as intended. The Germans had effectively no corps or army reconnaissance forces. They operated entirely effectively without them.

At Cassino there had been two good examples of British infantry battalions infiltrating in the advance and attack, seizing objectives surreptitiously and for little loss. In Normandy some junior officers stumbled onto the tactic.[19] There are few good examples of it being done deliberately. One was Canadian. Conversely the Germans were quite good at it. It didn't generally succeed against a well-integrated defensive position. It could, nevertheless, provide a low-cost route to low-level tactical success; from there to exploitation; and on from there to higher-level objectives. Of the major armies, the British were probably the worst at that. They had not trained for it.

The British were not good at handling all-arms armoured formations. There had been no-one to teach them. In the dark days after Dunkirk that role had fallen to Alanbrooke.[20] Although highly capable, Alanbrooke was a gunner with no actual experience of armoured warfare.

Allied defensive tactics were good. There were virtually no instances of German attacks penetrating beyond British forward companies. German attacks rarely prompted Allied divisional-level, let alone corps-level, counterattacks. That was partly due to massive artillery fire support in defence, but also to good minor unit tactics. Arcs of fire *were* interlocked (reflecting the earlier remarks about Bren guns). Antitank weapons *did* cover likely tank approaches. Mortars and artillery were available when needed. Reserves, with tanks, were on hand. Allied defences were generally more deeply echeloned than German: effective, but less efficient. That is, more troops were required across a given frontage.

German units were generally more effective tactically than Allied units. Some were poor. As the war progressed German quality tended to drop. British infantry fighting a German parachute regiment in December 1944 were surprised to find that the Germans surrendered fairly quickly. They then discovered that the regiment had only been formed in September.

Training and experience matter. The British 3rd Division had practiced amphibious landings about a dozen times. Some of those rehearsals were in the ships it would use on D-Day. Its landing plan was simple. Its landing went very well. Most British divisions had not fought since 1940, but had been training for four years or more. By the end of the campaign, most had fought for several weeks. So had many American divisions. One British brigade conducted an assault river crossing on each of three successive nights in

19 See, for example, Sidney Jary, *18 Platoon.* (Bristol: Sidney Jary Ltd. 3rd Edition, 1994), passim.

20 Alex Danchev and Daniel Todman eds, *War Diaries 1939-1945. Field Marshal Lord Alanbrooke.* (London: Phoenix Press, 2002), 129-135.

Normandy. One engineer company built five bridges, on five different rivers, in seven days. Given that much experience, units and formations became quite proficient. Only two more British divisions were sent to northwest Europe thereafter. One came from Italy. Allied divisions that had fought through Normandy generally went on to fight well. American divisions which came ashore after Normandy were often less good.

Post-war analysis by the American Colonel Trevor Dupuy and his colleagues compared British and American formations fighting the Germans in Italy. The analysis indicates that the US Army was far better at killing, wounding and capturing Germans than the British were. For example, in successful attacks the Americans inflicted 1.125 German casualties per American casualty. For the British the figure was 0.685 to one. Clearly the Americans were more effective than the British.[21]

Or were they? There is no reason to doubt Dupuy's statistics. However, effectiveness is the ability to achieve outcomes. By that measure the two armies were roughly equal. Americans succeeded in 51 per cent of their attacks. The British succeeded in 47 per cent. However, casualty ratios are measures of *efficiency*: the ratio of outputs to inputs. So the Americans were clearly more *efficient;* at causing casualties.

There is also a third measure: economy. Economy relates to minimising inputs (or reducing costs). In this sense a casualty is a cost. Each casualty means that a replacement has to be provided. Regrettably the Americans score far less well in that sense. When their attacks succeeded, Americans took 65 per cent more casualties than the British. Where they failed, they took 91 per cent more casualties. Over the 70 American attacks considered, that would equate to about 9,500 *more* casualties than the British. That means about 1,900 more dead. (The Americans suffered 21,616 casualties in 70 attacks. The British suffered 8,461 casualties in 49 attacks.)

So: the evidence is that the British and American armies were, in practice, equally effective. That is, they achieved their objectives in a similar proportion of the battles studied. The Americans were appreciably better at causing German casualties. The British were *much* better at avoiding casualties, by almost two to one. Two other statistics are telling. Firstly, the British planned their attacks with a greater numerical superiority over the

---

21 Christopher A Lawrence, *War by Numbers. Understanding Conventional Combat.* (Lincoln, Nebraska: Potomac Books, 2017). 22 and passim.

defenders. That is, almost exactly three to one on average. The Americans averaged just 1.80 to one. Secondly, when an attack failed, the British closed it down with very few casualties. As previously mentioned, an American attack that failed suffered 91 per cent more casualties than a British attack that failed. An American attack that succeeded took a 157 per cent more casualties than a British attack that failed.

The Germans also gave up more easily when fighting the British. When the Americans attacked successfully, the Germans suffered an average of 397 casualties before withdrawing. When the British attacked successfully, however, the Germans withdrew after suffering just 146 casualties. Clearly something was persuading the Germans to desist far earlier, in terms of casualties suffered.

British and Dominion commanders, including Montgomery, repeatedly said that American units didn't know how to fight. That clearly wasn't right. Compared to the British, American units were very good at killing, wounding and capturing Germans. They were just as good at winning battles, and perhaps slightly better. The Americans just lost far too many people in the process. Montgomery and his colleagues probably didn't know the details. He may also have been commenting on relatively inexperienced American divisions, newly arrived in Normandy.[22]

There is another highly revealing set of statistics. In 1944 US forces suffered 18,038 hospitalizations due to Immersion Foot (or 'trench foot') in northwest Europe.[23, 24] At the same time the British and Canadians suffered 39. Not 39,000; or 39 per thousand; but 39.[25] Trench Foot is largely preventable. French and British troops had learnt how to prevent it during the Great War. It is not good enough to blame poor-quality issue boots or socks. Poor boots and socks reflect an army that hadn't learnt that lesson. These are not reflections on poor soldiers, nor a poor army. They reflect an army that had not learnt some of the lessons of the Great War.

---

22 It has not been possible to make a similar comparison of American and British formations in Normandy, because the German casualty records were lost. Chris Lawrence, personal communication.

23 *A Report on Trench Foot and Cold Injuries in the European Theatre of Operations*. Octa C Leigh, US Army Medical Corps, 1946.

24 17,500 cases to 12 December 1944. *Time Magazine*, 1 January 1945.

25 There is a discrepancy of five cases, to a total of 44. That is probably due to the reporting date used. See also *The Administrative History of the Operations of 21 Army Group*, 88 and Appendix N.

British and American infantry casualties in northwest Europe show a fairly consistent pattern. Infantry casualties constitute about 70 per cent of all army casualties. Units that landed relatively early in Normandy typically suffered 100-120 per cent casualties over the campaign. Typically about 20-25 per cent of the original battalion would still be present at the end of the war. One British infantry company (of 120 all ranks) had 16 still with the colours, but not necessarily with the company, at the end. Infantry officers were twice as likely to die as their soldiers. In British tank battalions about 30 per cent of all ranks became casualties. 100-120 per cent of the tanks were damaged by enemy action. About 90 per cent of the officers were killed or wounded.

Medical statistics show a remarkably consistent ratio of between three and five wounded to one killed in action. The survival rate in field hospitals exceeded 98 per cent by 1944.[26] In Europe and North Africa one British or American soldier was admitted to hospital as a psychiatric case per 2.4 – 2.7 wounded; except for the British Army in northwest Europe. There, the rate was less than half that.[27] The reasons were, essentially, the lessons of the Great War. Psychiatric casualty rates for western armies in the Far East were generally much higher.

Medical statistics for Bomber Command show a different pattern. Officer aircrew hospitalization rates for medical (as opposed to surgical) conditions mirror the rate of wounding quite closely. Wounding almost invariably happened whilst flying, so wounding rates can be taken to be a good measure of combat activity. The pattern for NCO aircrew is different. Medical hospitalizations considerably exceeded the rate of wounding. They increased rapidly after 1940. Station-level evidence suggests that some disorders (hence medical admissions) were being used as a cover for apparent exhaustion. If that is so, why were officers not so prone to exhaustion? They flew the same missions in the same aircraft.

Another piece of evidence comes from disciplinary evidence. Senior RAF officers objected to the practice of unit and station commanders simply giving aircrew time off when they were not flying; rather than training, organised sport and so on. The evidence suggests large numbers

---

26 John Ellis, *The Sharp End of War. The Fighting Man in World War II.* (London: David and Charles, 1980), 169.

27 *The Administrative History of the Operations of 21 Army Group*, Appendix L3.

of undertrained, under-employed NCOs losing their motivation from 1941 onwards and increasingly reporting sick. It is related to a wider debate about 'lack of moral fibre' in aircrew. It suggests weak leadership.

There has been considerable, and often sensational, discussion of poor morale in various forces at various times. In autumn 1944 Eisenhower reported to Marshall that there were about 100,000 American 'stragglers' absent from their units in northwest Europe. The important question, however, is brutally simple. Did such 'poor morale' result in units being below strength; failing to achieve missions; combat refusal; desertion; or unauthorised surrender? If not, the issue is perhaps historically or sociologically interesting. Tactically it is relatively unimportant.

Some western army unit and formation commanders failed, generally the first time they fought. Others rose very rapidly and successfully through the ranks. Possibly a larger number rose rapidly and then failed. A review of British armoured division commanders suggests that the failures tended to come from socially selective regiments: Guards, cavalry and rifle regiments.

Dummy tanks and aircraft were commonly deployed as decoys. The Luftwaffe simulated burning cities in order to decoy RAF night bombers. The RAF and the Kriegsmarine had been intercepting tactical radio traffic since the mid-1930s. British ships monitored enemy radio communications to give early warning of air raids. By the middle of the war they could detect large raids, using radar, out to about 60 nautical miles. Radio intercept was generally better than the opposition's COMSEC.

Communications discipline was poor in many armed forces. In the Western Desert, the officers of 4th Indian Infantry Division were confident that Rommel's SIGINT service could not analyse radio traffic in Urdu. They were right. However, the ability to identify and locate that one division quickly and repeatedly, and place it in the order of battle, was a gift to the Germans. Similarly the Japanese Army did not believe that US forces could read Japanese writing. They were wrong. The Japanese were careless about document security, which was a considerable advantage to the Americans.

Failures could have serious consequences. The British aircraft carrier HMS *Glorious* was sunk off Norway in 1940 because of a SIGINT failure. It was not a failure of detection, interception, nor analysis; but of dissemination. *Glorious* was not told that enemy warships were known to be in the area. At roughly the same time, the Allies' ability to read German signals (codenamed 'Ultra', mostly produced by the Enigma coding device) failed for the first ten, critical, days of the French campaign.

Successes were also valuable. SIGINT allowed Royal Navy planners to predict which deep water passages would be used during the 'channel dash' of February 1942. The channels were quickly mined. Both *Scharnhorst* and *Gneisenau* were damaged as a result. On 17 May 1944 the German 1st Parachute Division declined to withdraw from Cassino. It had to be ordered to do so, by radio, by Field Marshal Albert Kesselring (the German commander in chief) in person. The Poles intercepted the order.

We can now make some high-level observations about warfare at the tactical level in the Second World War. In a sense, not much changed between 1939 and 1945. Several of the larger Allied warships fought throughout the war. Messerschmitt Me 109s and Spitfires flew both at the beginning and at the end. Panzer IV tanks, 25pdr field guns and Bren guns equipped their armies throughout the war. British and German uniforms and helmets generally did not change. Tactics were broadly the same in many cases.

Unlike the Great War, tactics did not change much from the summer of 1940. Most of the tactics that air forces, navies and armies used were demonstrated before Dunkirk. If they had not, they had at least been thought about. In 1939 most armed forces had a broad idea of what to do. Some were much better than others at actually doing it. By 1945, they had all become better at 'actually doing it'. What the Red Army was 'actually doing' was still quite crude. It had, however, become good at doing simple things well, quite quickly, and taking huge casualties in the process. It was still doing that when it captured Berlin.

The main change to tactics was, effectively, that they were much better done. By 1945 western armies integrated their tactics down to platoon level for tanks and infantry; to company level for mortars, artillery and engineers; and brigade or perhaps battalion level for air support. They were, typically, less likely to attack frontally. They used indirect fire support better. Units and formations could (and did) plan and execute quite sophisticated operations, such as night assault river crossings, in a few hours. British infantry platoons were expected to be attacking within five minutes of the enemy being located. Battalions given four hours' notice of an attack were expected to be 100 per cent successful. Similar considerations applied to naval and air force tactics. Integration went up as well as down. Tactical air forces were integrated at theatre level, but also right down to division or brigade level.

The weapons and platforms had actually changed significantly. Spitfires and Me 109s were no longer 'state of the art'. Many aircraft now carried electromechanical or electronic equipment: radars, gunsights, bombsights, sensors, jammers, receivers and radios. Ships were beginning to bristle with antennae: communicating, navigating, detecting, sensing, directing, jamming. Some of the smallest landing craft had navigation radar. Even the humble antiaircraft shell could now be radar-fused. Much of this was secret, not discussed openly, and was neither obvious nor remarked upon.

Tanks had roughly doubled in weight, from typically 15-20 tons to 30-45. Tank guns' calibres were generally 75mm or larger, double the 37 or 40mm of 1939. They could penetrate armour about six times thicker. Aircraft were generally not much faster, but bomb loads for heavy bombers had more than tripled.

Some weapons and equipment had not changed. As in the Great War, the principal field guns had not. Writ large, there *had* been considerable development. It was not so much in look, or name, as in performance. Similarly with tactics.

Several major urban operations (such as the attacks on Le Havre, Boulogne and Calais) were highly successful. The casualty statistics were highly favourable to the attackers. That suggests that much of the received wisdom about urban operations is deeply flawed.

Infiltration, using covered approaches to find and attack weak spots, had become the norm. That is: for the infantry of western armies; at section, platoon and perhaps company level. It was sometimes used at battalion level. Infiltration might be opportunistic. Alternatively, it was sometimes used in deliberate, planned night operations. German, and some American, armoured units could infiltrate. Others couldn't, or generally didn't. Infiltration was often more successful than more deliberate methods. It typically resulted in fewer casualties.

On land, the best armies could infiltrate, attack and then exploit. In particular they could translate a break-in into a breakthrough and then an exploitation. The Wehrmacht used the expression 'feel with the fingers; punch with the fist'. One might say that the British never learnt to do that. One could argue that Montgomery never really tried.

Air support to naval forces could be devastating. Close air support to land forces, and interdiction, could make a major impact. That might be on the battlefield, across a theatre, and throughout a campaign. To an extent the Bartholomew report was right. By 1944 the RAF were largely doing what the Army had expected in 1940. Very close cooperation was required to make that work in a timely manner.

Nonetheless, air attack is raiding. Whether in Burma or over Berlin, at Mortain or during Market Garden, aircraft attacked and flew home. That is raiding. Much of the value of any raid lies in surprise. All damage caused by weapons and munitions is tactical, and such damage can often be repaired or replaced, given time. Surprise is transient. The effect of surprise, and raids, wears off if not exploited rapidly.

Once again, we have seen that historians have missed things several times.

Finally, don't confuse effectiveness with efficiency or economy. The business of armed forces is all about outcome. That is, winning battles and engagements; hence campaigns; and hence wars. (Military) effectiveness relates to outcomes. However sometimes the best balance of efficiency and economy, required to maximise effectiveness, is not obvious. Be that as it may, western countries *will* put a high price on the ability to bring soldiers, sailors and airmen home at the end of the war. That relates to one important form of economy: keeping casualty lists short.

9

1945-1975

Chapter 9 considers the 30 years immediately after the Second World War. With some major exceptions, the period from 1945 to 1975 was not marked by great technological change. Most of those changes were underway before the end of the Second World War. So, in Chapter 9 we will focus on global geostrategic circumstances and then four major wars (Korea, Vietnam, the Six Day and the Yom Kippur Wars). We will then examine the phenomenon of insurgency as applicable to this period. One of the few major technological developments was the introduction of nuclear weapons. We will consider that briefly before making observations and deductions.

We can identify four major geostrategic issues in this period. They were: the neutralisation of Germany and Japan; the prominence of Communism; the rise of multinational organisations; and the decline of colonialism.

It is difficult to imagine what the world would have looked like in 1950 with a powerful Germany and a powerful Japan. Because they had surrendered unconditionally, there had been no repeat of the Paris Peace Conference. In practice the two countries were neutralised. They did not, and largely still do not, play a major role in international relations. Their economic power is much larger than the part they play in world affairs. That is incredibly different from the situation in 1914 or 1939. In 1945 both countries were occupied. Both had new political institutions imposed on them. Unusually, both German and Japanese militarism were effectively neutralised. So, as a generalisation, to impose lasting political change on a country, it must be occupied and new political institutions imposed on it. That is a generalisation. It has many implications. It has many caveats. In broad terms, however, it seems valid.

Looking at Communism in retrospect, we can say firstly that the Soviets imposed Communist governments on several countries in central and eastern Europe. Those countries universally rejected Communist governments after 1989, as did most of the new states formed within the USSR. Secondly, the Soviet Union supported a large number of insurgent groups elsewhere, but very few of them formed Communist or Marxist governments after independence. Thirdly, the forces of the Chinese Communist Party won the civil war in China in 1949. Communist regimes were imposed there and subsequently in North Korea, Vietnam and Laos. They still hold power in all four countries.

None of those countries have held free and fair multi-party elections. Nor did the various eastern European countries before the fall of the Iron Curtain in 1989. Thus we can say that Marxist-Leninist groups were successful at seizing power in a number of countries after 1917. They were also (and in some cases are) quite successful in retaining power, mostly through various degrees of repression. Politics, of itself, does not concern us in this book. The use of collective violence for political purposes does. We can conclude that repression *can* succeed for decades. In some cases it continues to do so.

In two large countries wracked by social upheaval (Russia in 1917 and China, which collapsed after the failure of the imperial government in 1911), well-organised and vicious Marxists or Leninists seized power. They then fostered, supported or imposed seizures of power in several adjoining countries. The core Marxist-Leninist groups (often called 'cadres') were quite effective. From that *military* perspective, Communism worked. The evidence does not tell us much (if anything) more about Communism which is relevant to the study of war and warfare.

Many international organisations have been created since the Second World War. The largest and most successful is the United Nations (UN). It originated in the European governments in exile in London in 1941. It grew significantly after the United States joined in 1942. The UN became truly global after the Second World War. It is now hard to imagine a world without it. It is similarly difficult to imagine how the world functioned before it. In practice, all but the most flagrant violations of international order are referred to the UN before the outbreak of war; and often even in those cases. So the UN is, in effect, an important factor in strategic calculation for any county.

There are now many other international organisations. Today the creation of another is not necessarily remarkable, except (presumably) to its members. Many are regional. Some have collective security arrangements. Unusually, some involve standing armed forces. During the postwar period the most significant were NATO and the Warsaw Pact. Perhaps surprisingly, their coexistence underpinned decades of uneasy peace in Europe. Under that 'peace' the western democracies and the Soviet Union sought advantage elsewhere. That largely meant in what was called the 'Third World'. Those conditions prompted a broad 'non-aligned' movement of states, in the UN and elsewhere. After 1947 India was a major figure in the non-aligned movement.

Turning to decolonisation: we should remember that it was underway before 1939. Most South American countries had become independent in the nineteenth-century. Cuba became independent from Spain in 1902. Britain effectively decolonised its white dominions (Canada, Australia, New Zealand and South Africa) in 1926. It enacted the legislation to decolonise India (hence also Pakistan, Sri Lanka, Bangladesh and Burma) in 1935.[1] Lebanon became independent from France in 1943.

Decolonisation was largely complete by 1975, which defines the scope of this chapter. About 80 former colonies are now independent and members of the UN. Strategically, one should remember that *every* country that joined the UN agreed to the principle of self-determination.[2] Self-determination implies decolonisation (where requested).[3] *All* the former colonial powers agreed to decolonise, by the act of joining the UN. Decolonisation was specifically endorsed by a UN declaration in 1960.[4]

Thus, in conflict after conflict, the so-called 'liberators' were largely pushing against an open door. That is usually lost in the noise of academic debate. There are three useful examples. Firstly, Fidel Castro portrayed himself as the darling of the anti-colonialist movement. In 1952, however, he took over a county which had actually been independent for 50 years. Secondly, the Kenyan Emergency delayed independence for Kenya. That was not because Britain refused to grant it, but because Britain would not abandon a colony during a major insurgency. Once the (Mau Mau insurgency was resolved, independence was granted. Thirdly, when the white minority in Rhodesia declared independence unilaterally in 1964, Britain refused to recognise it. 'White minority rule' did not constitute

---

1 The Government of India Act, 1935. Self-government was granted under its provisions in 1947. The so-called 'Indian Independence Act' of 1947 actually partitioned India into two Dominions. They became legally independent of Britain in 1950 and 1956 respectively.

2 Article 1(2) of the Charter of the United Nations.

3 Several small territories remain colonies, or similar, in accordance with the wishes of their populations. The Falkland Islands and Gibraltar are examples.

4 United Nations Resolution 1514 of 14 December 1960.

'self-determination'. Britain granted independence when elections were held on the basis of 'one person, one vote' (in 1979 and 1980). Rhodesia then became Zimbabwe. Those are the facts. However, one immediately hears the howls of protest from among the International Relations community.

Not least due to academic misrepresentation, it is easy to forget how many of the conflicts of the period arose in colonies or former colonies. Most, if not all, were related to gaining (or granting) independence. See Figure 9-1.

Modern Name	Former Colonial or Mandated Power	Year Granted	Remarks
(a)	(b)	(c)	(d)
Laos	France	1953	Civil war related to Vietnam 1953-75 (see below).
Vietnam		1954	a. North and South Vietnamese independence in 1954. b. South Vietnam overrun in 1975. c. Invaded by China in 1979.
Cambodia		1954	Invaded by Vietnam in 1978.
Algeria		1962	Insurgency from 1945 resulted in the Algiers Putsch of 1961, then independence in 1962.
Korea	Japan	(1945)	Resulted in the Korean War.
Indonesia		1949	Insurgency 1945-9. See Timor-Leste.
Guinea-Bissau	Portugal	1973-4	Insurrection from 1956.
Angola		1975	Insurgency from 1961. Civil war from 1975 to 1990. Cuban, South African, US and Soviet support. See Namibia.

continued over

Mozambique		1975	Insurgency from 1964. Independence led to removal of support for Rhodesia and contributed to Zimbabwean independence.
Timor-Leste		1975 & 1999	Occupied by Indonesia in 1975. Independence from Indonesia in 1999.
Namibia	South Africa	1990	Insurgency from 1966. Independence granted once Angolan conflict resolved (see above).
India, Pakistan and Bangladesh	United Kingdom	(1947)	Indo-Pakistani wars in 1965, 1971 and 1999. See also Sri Lanka.
Sri Lanka		1947	Subsequent Tamil insurgency.
Israel		1948	Arab-Israeli wars 1948, 1956, 1967, 1973, 1982 (and conflict continues).
Egypt		1952	Then Arab-Israeli wars (see Israel).
Malaya		1957	a. Communist insurgency from 1947. b. Separation of Brunei, and then Singapore in 1963.
Cyprus		1960	a. Insurgency by EOKA from 1955. b. Continued inter-factional conflict until 1974. c. Partition from 1974 (continues).
Kenya		1963	Mau Mau insurgency from 1952 to 1960.
Aden		1968	Subversion and then insurgency from 1956.
(Rhodesia) Zimbabwe		1980	Britain had refused to grant Rhodesia independence until majority rule was agreed. Subsequent black African insurgency against the white Rhodesian government.
Northern Ireland		(1922)	Republican insurgency from 1969 to 1976. Then terrorist campaign to 2006.

Figure 9-1: Selected Conflicts Arising from Decolonisation

The list provides several clear examples of how the Second World War left unfinished business. Korea, Vietnam and a homeland for the Jews (in what became Israel) are cases in point. The issue of a homeland for the Jews had emerged before the Great War. Even today (2024) it can scarcely be considered 'finished business'. As previously mentioned, 'decolonisation' for India was provided for in 1935. More broadly, there is always unfinished business. Wars can also create new political business, which may then run on for decades.

The fact that the United States fought both the Korean and (Second) Vietnam Wars clouds those countries' origins as former colonial territories. Korea was effectively annexed by Japan in 1910. At the end of the Second World War it was divided into American and Soviet zones.

By 1950 South Korea had an army of 200,000 men trained to guard the border and for counterinsurgency. By the summer of 1950 it had reduced the number of communist insurgents operating in the south from 5,000 to about 1,000. In 1945 the USSR had appointed Kim Il-Sung, a former Red Army officer, to lead the (north) Korean Communist Party. In late 1949 Mao, Kim and Stalin had held talks in Moscow. China returned 50-70,000 Korean soldiers, who had been fighting against the Chinese nationalists, to North Korea. The North Korean People's Army reached a strength of 11 divisions.

North Korea invaded in June 1950. 85 per cent of South Korea was overrun within a month. American air forces slowed the advance, creating time for the formation of a defensive perimeter around the southern city of Pusan. MacArthur initiated an inspired counteroffensive with an amphibious landing at Inchon in September. That threatened to cut the North Korean Army off. American-led UN forces exploited rapidly north towards the Chinese border.

China then intervened. About 150,000 Chinese soldiers crossed the Yalu River into Korea in November. Much of their air support was Soviet. After much fighting the front line stabilised roughly along the 39th Parallel in July 1951. The war came to an end, after protracted negotiations and further fighting, in 1953.

US air forces had switched to attacking targets in northern Korea. By the end of the war practically every town in the north had been severely damaged. US air forces dropped 150,000 tons more bombs on Korea than they had in the Pacific during the Second World War. The United States Air Force (USAF), which had become independent in 1947, was justifying its existence. Yet American air forces never achieved air superiority, mostly due to enemy ground-based air defences. In addition, it is not at all clear that aerial attack *could* coerce North Korea.[5] Put simply, there was little of value to bomb.

Vietnam was one of three colonies in French Indochina. The others were Laos and Cambodia. Vietnam had been occupied by the Japanese, liberated itself, and was then re-occupied by the French. An insurgency developed, led by the communist Viet Minh. Initially the Viet Minh were not the only faction, nor the most powerful. However they brutally suppressed the opposition, including Trotskyites in their own movement. They broke or substantially weakened all rival anti-colonial forces.

The French fought against Vietnamese communist insurgents for nine years, under several governments and several commanders-in-chief. In 1954 General Henri Navarre developed a plan to defeat the North Vietnamese Army (NVA) in northern Vietnam. He dropped a division of paratroops, and then built an airfield, at Dien Bien Phu. After five months of preparation and a seven-week battle, the paratroopers surrendered. The French could then no longer defend northern Vietnam, despite massive American financial support. Vietnam was divided along the 17th Parallel.

By the early 1960s large numbers of communist insurgents were operating in South Vietnam. The Strategic Hamlet Program, initiated in 1962, was inspired by similar policies in Malaya and Kenya. It was very poorly executed and abandoned in 1964. The insurgency grew. A massive American troop build-up started after the Gulf of Tonkin incident, also in 1964. American troop levels increased from about 16,000 men to 549,000 by 1969.

America's war aims were limited to defending South Vietnam. Could America have done that? Of course it could. Doing so might have required even higher levels of commitment and possibly even the use of nuclear weapons. But it did not. It failed. President Lyndon Johnson tacitly admitted failure when he ordered a halt to the bombing and declined to run for re-election in 1968. His successor, President Richard Nixon, withdrew American forces. South Vietnam was overrun in 1975. However, at much the same time, Nixon managed to engage with Communist China for the first time. He also persuaded the USSR to sign the first Strategic Arms

5 Gray, *Airpower for Strategic Effect*, 167.

Limitation Treaty. It might reasonably be said that Nixon accepted one strategic failure in the process of achieving two major strategic successes.

America lost about 58,000 dead in Vietnam. It had lost about 36,500 dead in Korea. At its peak, in 1968, the Vietnam War cost an average of 46 dead per day. One officer was killed for every 7.81 soldiers: an unusually high proportion.

That analysis overlooks the part played by the North Vietnamese. They realised that they could not defeat American forces in battle. Therefore they developed a long-term delaying strategy, linked to an information campaign in America and elsewhere. US domestic support for involvement in Vietnam fell. Put simply, North Vietnamese strategy was successful.

The American operational approach was initially based on attrition. However, US forces could not prevent the North Vietnamese remaining in the south but avoiding contact (and hence attrition). Attrition does not work unless the enemy is fixed and then forced to accept unsustainable losses. It didn't work in Vietnam.

In the Battle of Hue in 1968, American and South Vietnamese forces lost 668 killed in action. The Viet Cong lost at least 1,042, and possibly as many as 5,000, dead. The great majority of casualties on both sides were suffered when the American and South Vietnamese counterattacked to recapture the city. Hence, yet again, in a major urban operation the defenders (the Viet Cong at that stage) took more casualties than the defenders.

The air campaign was initially a 'controlled escalation'. It soon ran out of targets. Thereafter the USAF conducted several thousand armed reconnaissance (that is, airborne 'search and destroy') sorties every month. America dropped 211,000 tons more bombs in Vietnam than in Korea. Napalm usage increased from 32,000 to 388,000 tons. 20 million gallons of defoliants were sprayed, much of it for crop destruction. More than 20 per cent of the crop-growing area of Vietnam was sprayed every year for nine years.

In December 1972, under Operation Linebacker II, about half of all the B52 bombers in the USAF were used to bomb Hanoi. The objective was to force North Vietnam to sign an agreement which it had agreed to sign several months before. Aircraft losses were very high. In practice, America then signed concessions. Air forces had been misused in a would-be coercive and brute force bombing campaign that made little strategic sense.[6] One is

6 Ibid, 181.

struck by the disparity between the immense effort as input and little, if any, strategic reward as output.[7]

It has been claimed that America won every (land) battle in Vietnam. That might be true; it probably wasn't; but it was irrelevant. There was effectively no operational planning. America largely fought battles that did not contribute to strategic success. Its government had perhaps been misled by theorists such as Thomas Schelling, Robert Osgood and Bernard Brodie. They stressed ideas such as 'limited war' which stressed geographic limits, constraints on levels of violence, limiting means and managing escalation. Such measures have value, but they should not drive the conduct of a war. Those measures qualify the means and the ways. They tend to overlook the ends: the goals sought. The effect of such theory was to overstress the 'how' to the detriment of 'why?'[8]

On 23 May 1967 Egypt closed the Straits of Tiran to Israeli shipping. Israel mobilised. On 5 June it attacked, in what became the Six Day War. The Israeli Air Force (IAF) destroyed 393 Arab aircraft on the ground, most of them on the first morning. Three small Israeli armoured divisions struck into the Sinai. Eight Israeli brigades faced eight Egyptian divisions. Within 48 hours the Israelis broke through and forced the Egyptians to evacuate the Sinai. On 6 June, eight other Israeli brigades attacked into the West Bank. They drove the Jordanian Army out in two days. The 8th Israeli Armoured Brigade drove 350kms from the Sinai to the Golan and took part in the assault on the Golan Heights on 9 June. It was joined by two armoured brigades which had just fought in the West Bank. A ceasefire was declared on the 10th. The Israeli Defence Forces (IDF) held the West Bank, the Golan Heights and all of the Sinai.

It was a crushing defeat for the Arab armies. It was a brilliant example of high-tempo combined-arms operations. Destroying most of the Arab air forces on the ground gained air superiority. That allowed the IAF to support ground forces almost wherever required. Israeli armoured forces conducted daring, high-speed attacks deep into enemy positions. That created confusion, panic, retreat and rout. On occasion Israeli infantry forces attacked fast, hard, and typically at night; with similar results. The

7 Ibid, 183.

8 Donald Stoker, *Why America Loses Wars. Limited War and US Strategy from the Korean War to the Present.* (Cambridge: Cambridge University Press, 2019), 33-4.

main observation is that, once again, integrated land-air operations, led by armoured formations, can have swift and decisive operational effect.

Egypt and its allies were then re-trained and re-equipped by the Soviet Union. Arab strategic and operational goals for what became the Yom Kippur War in October 1973 were modest. The Egyptian Army would seize crossings of the Suez Canal and occupy a deep zone in the Sinai. Syria would seize the Golan Heights down to the Jordan River. Both would provoke Israeli counterattacks. They would be beaten off, forcing Israel to make political and territorial concessions.

Israel was largely taken by surprise. Front-line troops were at high alert, but reservists were not mobilised. Egyptian engineers created 60 crossings of the Canal in the first 12 hours. Five Egyptian infantry divisions, followed by armoured formations, crossed the Canal and advanced into the Sinai. Israeli armoured counterattacks were beaten off by massed antitank guided weapons (ATGW). The IAF was driven off by surface-to-air guided missiles (SAMs). The IDF went over to defence in the Sinai whilst dealing with the more immediate threat on the Golan.

There, 15 Syrian brigades were initially met by one IDF armoured brigade and two infantry battalions. The IDF's defence was epic: one tank in the Israeli 53rd Tank Battalion destroyed 35 Syrian tanks. The Syrians did not cross the main Israeli antitank ditch in enough places. No Syrian infantry fighting vehicles penetrated beyond the forward IDF infantry strongpoints. Unsupported Syrian tank units were then destroyed by Israeli counterattacks. The IDF went over to the attack. Jordanian and Iraqi formations were committed. The Jordanians were well led. The Iraqis were cut up badly. Both were defeated.

The Israeli counteroffensive in the Sinai broke through Egyptian defences, defeated their armoured reserves and crossed the Canal. The Egyptians quickly withdrew from the Sinai. In 17 days the IDF defeated the Syrian Army and its allies, recovered the Golan, and threw the Egyptian Army back out of the Sinai.

With few exceptions, the armies used Second World War tactics. Israeli M48s, up-gunned Shermans and Centurions had fought post-war Soviet tanks, and occasionally Jordanian Centurions. Success generally went to better gunnery, not better tanks. Little that the Israelis or Arabs did would have surprised Guderian, Patton, Bradley or Montgomery.

Earlier that year the commander of the IDF Northern Command had seen Syrians using ATGW in skirmishes. He began to develop countermeasures. That included increasing the number of mortars in tank battalions. He was too late. Similarly, the IAF initially had difficulty in countering Soviet-built SAMs. After a few days' fighting Arab SAM sites were being destroyed through a combination of cluster bombs dropped from very low level, anti-radiation missiles, and long-range artillery fire.

Once again, integrated land-air operations by armoured formations had swift and decisive operational effect. Arab generals were often competent and their soldiers brave. Their junior and middle-ranking officers, however, followed orders slavishly and lacked initiative. Israeli armoured formations were, genuinely, commanded on the move. Arab forces had to be commanded to move. However, once again, Israel achieved operational success but no lasting strategic settlement.

Many of the wars of decolonisation started as uprisings; that is, insurgencies. The colonial power generally responded with counterinsurgency campaigns. What did that mean strategically? As previously discussed, in practically every case the colonial power was going to decolonise. It did. So it achieved its broad strategic goal. The questions were 'how' and 'when?' So, what did strategic success mean? To either party? Can both sides 'win'? Is victory anything other than a declaratory political artefact? In many cases the colonial power had no interest in *declaring* victory. That may simply be because it wished to stay on good terms with its former colony.

Several insurgent groups were nominally Marxist or Communist. As previously mentioned, few of them subsequently formed communist governments. On reflection, Soviet attempts to gain influence (or simply cause trouble) coincided with the insurgents' own interests. Much was made of an apparent global spread of Communism. It was overblown.

There were several successful counterinsurgency campaigns. That is to say, campaigns in which the colonial power achieved an outcome which met its strategic purposes. Where they succeeded, they often showed restraint in the use of force. That is not quite the same as 'the minimum use of force'. They also generally sought the support of the majority of the indigenous population. That is, broadly, 'hearts and minds'.

In May 1945 the French Navy used a six-inch gun cruiser to shell villages in Algeria (which was, at the time, a department of France). That is scarcely restraint in the use of force. It was not calculated to gain the support of the population. It is perhaps not surprising that Algeria was independent by 1962. We will consider counterinsurgency in more detail in Chapter 12.

The Second World War ended, and the postwar period started, with the detonation of two atomic bombs. The following decades were dominated by the prospect of atomic and then nuclear war. That domination affected world politics, the popular media, and defence policy and spending. Yet nuclear war never happened.

Early plans for the British V-Bomber force of the late 1950s involved just one raid, by 100 bombers. They would fly at high level and high subsonic speed to bomb 100 cities in the Soviet Union. Each bomber had one free-fall nuclear bomb. 100 aircraft, 100 bombs, 100 cities. After the aircraft took off there was *one* 'go / no go' confirmation signal. There was *no* 'recall', by design. It was chilling. American and, presumably, Soviet plans were similar.

Such plans relied on perceptions of narrow technological superiority which quickly evaporated. For example, better Soviet missiles and interceptors rendered V-bomber tactics unworkable within a few years. What followed was a period of adaptation and evolution.

Nuclear devices became smaller and smaller. Eventually six-inch (or, rather, 155mm) artillery could fire nuclear shells. That meant that, for example, an enemy divisional or brigade headquarters could be a worthwhile target. Alternatively, a counterbattery task that would have required a brigade of heavy artillery in 1944 could now be undertaken with a single shell. Naturally the same applied to the enemy's use of nuclear weapons.

What can we observe or deduce from the immediate postwar period? In practice, for all the extensive theorising about nuclear warfighting, it has not happened since 1945. That is not to say that all that concern was pointless. The threat of the use of nuclear weapons forced all parties to think at great length about how to use them. They generally decided to avoid their use if possible, and encourage the enemy to avoid using them. Additionally, for all of the emphasis placed on air forces for their role in 'strategic nuclear warfare', the air forces supported the nuclear threat, not vice versa.[9] Similarly the development of ballistic missile submarines gave a highly credible, undetectable and dependable nuclear role to navies.

We should not think of nuclear warfighting as intrinsically irrational. It is entirely rational to seek to wage wars, if wage them one must, to win quickly and at little cost. Nuclear weapons offered that possibility. It is entirely rational, and entirely ethical, to seek to protect one's population and armed forces from the massive damage that nuclear weapons could cause. Rationally, that may extend to threatening massive damage in return. And, rationally, such threats must be credible. Once invented, nuclear weapons could not be un-invented. No-one has ever completely trusted an adversary with nuclear weapons to not use them. Subjectively, the use of nuclear weapons might be deeply *objectionable*. Considering how best to use them, however, is not *irrational*. The air marshals, admirals and generals were not mad.

Armed forces do not just fight. They manipulate the use, and the threat of the use, of collective armed violence. Much later one American politician pointed out that the United States uses nuclear weapons every day. To the extent that America *manages the threat* of the use of nuclear weapons on a daily basis, that is entirely true.

Put simply, air forces were shown to lack the power of decision; in Korea and Vietnam at least.[10] The West inevitably lost its atomic and then its nuclear monopolies. Nuclear warfighting, largely intended to be waged by air forces, didn't happen. The two attempts to win wars by bombing (Korea and Vietnam) didn't work. Conversely close air support and interdiction, integrated into fast-moving armoured operations, worked brilliantly well (witness the Six Day and Yom Kippur wars). Airpower theorists have criticised the role of the Israeli Air Force (IAF) as having been nothing more than a 'bombing contractor' for the Army. They do so partly to stress a more independent role for air forces. That seems mistaken. Not least, it ignores how brilliantly successful the IAF had been as a bombing contractor.

It would have been entirely possible for the United States to defeat North Korea in the 1950s. That might have required using atomic weapons, or mobilising on the scale of 1941-45. Yet, 70 years after the Korean War, there is no lasting political settlement. Why is that? We can conjecture

9 Gray, *Airpower for Strategic Effect*, 162.
10 Ibid, 182.

two things; both are informative. Firstly, America did not (and probably does not) think that the possible gain of 'winning' was worth the probable cost. Secondly, there is considerable advantage in permanently basing American forces close to (say) Vladivostok and Shanghai. Both aspects reflect strategic calculation.

Vietnam, however, tells us something else. It tells us very clearly that the operational level of war is critical. Winning every battle is, regrettably, worthless (and pointless) if it does not lead to winning the campaigns and hence the war.

We can also identify a continuing trend. Integrated air-land, armoured operations can win wars and do so very quickly. Conversely counterinsurgency can take a long time, and the fighting itself is rarely decisive.

The outcomes of the wars of decolonisation allow us to see very clearly that victory is a declaratory political artefact. Declaring victory can have political benefit. Conversely there may be reasons for *not* declaring victory. That may partly explain why veterans of some conflicts feel that their sacrifices were not acknowledged. It is possible that, strategically, it was not considered advantageous to do so. In such cases it might be that the least said, the greater the political benefit. That may seem cynical. Welcome to war.

Finally, strategic calculus may be wrong. In 1900 Britain had perhaps the third, fourth or fifth largest economy in the world. In the 1950s that was broadly unchanged. Yet Harold Macmillan's government convinced itself that its role was to manage strategic decline. By some measures, it was right. Britain had divested itself of its empire and was decolonising. Several countries now produced more coal, more steel and more cars than Great Britain.

But in terms of what had become cutting-edge, high technologies such as atomic weapons, military jet aircraft or computers in use, Britain remained one of the leading countries in the world. Its government had simply asked the wrong questions.

10

1975-2000

In considering the last quarter of the twentieth century, we shall again start by looking at the geopolitical situation and how it developed. We shall briefly consider military technology and operational concepts. We shall discuss the last decade of the Cold War (particularly on NATO's Central Front); then three specific conflicts; and finally the 1990s as a period. Those conflicts are: Northern Ireland; the Falkland Islands; and the First Gulf War of 1990-91. We shall then make observations and deductions.

To start with the geopolitical situation: by 1975 decolonisation was substantially complete. The Cold War, however, still had 14 years to run. It is, surprisingly, not clear why the Soviet Union then fell apart. One theory is that Soviet premier Mikhail Gorbachev simply made a mistake; perhaps one that he was led into by American President Ronald Reagan. Whatever the reason, the speed with which the Soviet Union disintegrated, and the eastern European communist countries fell away, was dramatic. So we can put a fairly precise date on the end of the Cold War. That is: late 1989.

The 1990s were several things. They were 'the decade of the Balkans', in which Europe and NATO tidied up the mess caused by the collapse of Yugoslavia. The 1990s were also a period of growing economic prosperity and democracy in South America and Africa. They saw the USA stepping slowly, hesitatingly and not always successfully into a period of superpower monopoly.

Remarkably, one thing that did not happen was any re-drawing of borders in Africa. For decades foreign relations experts had written that one of the continent's biggest problems was that Africa's borders had been drawn up by the former colonial powers. Some such experts still do. Yet in (say) 2020 almost all African borders were precisely where they had been in 1945.

Military technology in use developed rapidly, mostly due to Cold War rivalry. The Cold War may prove to have seen the fastest sustained rate of weapons development ever. By 1975 jet engines were used in most types of military aircraft, including helicopters. Solid-state electronics had replaced valve technology. As a result electronic systems were cheaper, more widely used and more reliable.

Almost all of the equipment in use had been built since 1945. It was surprisingly uniform in some ways. Many western air forces flew US Phantom jets. Almost every western country's tanks mounted the British L7 105mm gun. (The main exceptions were France, and Britain which had already moved on to the L11 120mm gun). Many western nations used the Belgian FN FAL assault rifle and MAG machinegun.[1] Warsaw Pact countries mostly used Soviet equipment, or license-built copies. Other nations generally chose between western and Warsaw Pact equipment.

One of the main applications of improved electronics was in guided weapons. They had been used *en masse* for the first time in the Yom Kippur war. The first major use of guided weapons at sea was during the Falklands Conflict of 1982. Guided weapons prompted some changes to low level tactics. Yet they also resulted in erroneous claims as to redundancy.

In 1973 critics announced, publicly and loudly, that the tank was dead. It wasn't. It still isn't. Similar claims several times, in relation to tanks and other weapon systems.

They tell us two things have been made. The first is that much military punditry is actually superficial and amateur. That is despite the apparent credentials of some of the pundits. The second thing is that weapons development tends swing between measure and countermeasure, as opponents seek technical advantage. Armed forces have typically made considerable investments in their major weapon systems. They will therefore go to considerable lengths to keep them viable as new threats develop.

For the US Army, the late 1970s was perhaps the beginning of what might be the world's first (or only) genuine revolution in military affairs ('RMA'). Alternatively, that RMA might not have happened.

By 1980 America was bringing five new weapon systems into service. They were the M270 Multi-Launch Rocket System, the AH-64 attack and UH-60 utility helicopters, the M1 Abrams tank and the M2-M3 Bradley Infantry Fighting Vehicle System. Perhaps more importantly, the Army also introduced a new operational concept, AirLand Battle. The concept

1 Known to British and Commonwealth armed forces as the SLR and the GPMG respectively.

sought to harness long-range sensors, weapons and computers in order to see, and strike, deep into the enemy's battlespace. That would support a sophisticated process of engaging the enemy's tactical and operational reserves. In turn that would allow planners to shape the contact battle, and hence win more easily at both tactical and operational levels. AirLand Battle was adapted, and adopted, by NATO as Follow-on Forces Attack (FOFA).

That was linked to the explicit recognition of the operational level of war. Staff colleges started to teach it and how to fight at it. New, long courses were introduced specifically for that purpose. However it is questionable whether that is a good use of the students' time, given how few of those students will ever operate at the campaign and theatre level.

The importance of decentralization of command was recognised in some western armies at much the same time. 'Mission Command' was introduced widely. At the time it was said that it would take a generation for mission command to become habitual and instinctive. Comparison of doctrinal statements, and analysis of actual behaviour, suggests that a generation later (that is, by 2020) that was an understatement. (By 2020 written doctrine had tended to swing towards the commander's duty to act, and away from trust and decentralisation.)

Turning to the Cold War: its focus was NATO's Central Front and, within that, the Inner German Border. Germany was divided into two armed camps. Unusually, two alliances with forward-deployed standing armed forces faced each other through decades of relative peace. Defence spending was at historically high levels. Armed forces exercised frequently and at large scale. Several corps- and divisional-level field training exercises took place every year. New equipment was introduced and upgraded regularly. For example, tanks might be upgraded after about ten years and replaced after 20. Thus, for example, the British Centurion entered service in 1945. It was replaced by the Chieftain in the mid-1960s, and then the Challenger in the mid-1980s.

As a result there was a high level of tactical competency. That was linked to a reasonable degree of continuity of practice from the Second World War. Inevitably some good lessons were forgotten. Most veterans of the Second World War had retired by 1975. Perhaps the abiding impression is that the post-war generation were never quite aware of what they knew. Thus they were not aware of what they forgot.[2]

More positively, the Yom Kippur War was analysed at length. Western armies replaced their antitank guns with guided weapons (ATGW). They also up-armoured their tanks with 'Chobham' compound armour to protect against enemy ATGW. The British Army revised its planned logistic consumption rates (and stocks) significantly. 155mm artillery ammunition scalings increased five-fold; mortar ammunition more than threefold; fuel provision two and a half times.

The US Army reorganised its divisions several times between 1945 and 1989. The most significant, and most flawed, was the 'Pentomic' structure introduced in 1957. It attempted to remove the brigade level of command. To do so it increased the span of command to five subordinates at both battalion and divisional level. The British 'Wide Horizon' trial of 1975 attempted something similar. Both attempts overlooked two issues. The first is that, under stress, the practical span of command *narrows*, to two or even less. The second is that in the heat and stress of the Second World War all armies' armoured divisions were, or became, very small. What was needed was not wider spans of command. It was smaller divisions.

By twenty-first century standards, air force doctrine was in its infancy. Colonel John Warden of the USAF was probably the father of modern air force doctrine. He was relatively unknown until the Gulf War of 1991. Air forces, and particularly NATO's Second and Fourth Tactical Air Forces on the Central Front, were nevertheless well aware of the importance of theatre-level interdiction. They had focussed on delaying the arrival of Soviet and Warsaw Pact theatre-level reinforcements well before AirLand Battle. Road and rail bridges over the Vistula and Oder Rivers would have been attacked frequently. Air forces, and particularly NATO air defence forces, were kept at high alert and high levels of proficiency.

NATO's sea lanes across the Atlantic were, potentially, highly vulnerable. The Soviet Navy's submarine force was enormous, as was its land-based naval aviation. Thus much of NATO's navies were focussed on antisubmarine warfare and air defence. There was little (if any) real surface threat. The main exception was in the Baltic, where Warsaw Pact amphibious

2 See Jim Storr, *Battlegroup! The Lessons of the Unfought Battles of the Cold War*. (Warwick: Helion and Company, 2021), passim.

forces posed a challenge. NATO navies, like its air forces, achieved a high degree of international interoperability. That was in part due to the existence of standing (multinational) naval forces such as the Standing Naval Force, Atlantic. On land there was virtually no interoperability below corps level.

Sectarian violence broke out in Northern Ireland in 1969. It soon developed into a republican, nationalist insurgency. That was largely defeated by a troop surge (Operation Motorman) in 1972. By 1976 the so-called 'Troubles' had evolved into a well-run terrorist campaign. In 1979 a terrorist bomb attack (at Warrenpoint) killed 18 soldiers. Developments in EW then effectively contained the bomb threat. Although the campaign ran on until 2006, republican terrorists never again killed more than 10 soldiers in any year. They rarely killed as many as five. The main republican terrorist organisation, the Provisional Irish Republican Army (PIRA), had been forced into a niche (described in Chapter 12).

PIRA had been effectively contained. After 1979 it deliberately maintained a strength of fewer than 300 active members. For security reasons it used a cell structure. It had, nonetheless, been penetrated. Its counterintelligence unit was infiltrated. As a result, PIRA units became incredibly conscious of threats to their security. That was understandable, and reached the point of paranoia. It seriously limited PIRA's ability to conduct terrorist attacks.

Terrorism is generally defined as the use of violence to inflict fear for political purposes. *To that extent,* PIRA was defeated. The Troubles were resolved through political negotiation in 2006.

Margaret Thatcher was the British PM from 1979 to 1991. Although emotionally engaged in the conflict, she made no political initiatives in Northern Ireland. Her successor, John Major, started a process which led to the Good Friday agreement of 2006. It was the longest campaign that the British Army ever fought. A total of 3,532 people died, including 705 soldiers and marines.

British armed forces almost always displayed restraint in the use of force. It was difficult to gain and retain the support of the Catholic, broadly nationalist element of the population. However, despite active propaganda and some repressive violence by PIRA, active support for Republican terrorists was limited. Government and Army information activities were initially woeful and inadequate. They slowly gained effectiveness.

Argentinian governments only draw attention to their claims of sovereignty over the Malvinas (Falkland) Islands in order to distract domestic political attention. That was the case when Argentina invaded the Islands in 1982. Britain's case was, and is, simple: self-determination. The Falklands Islanders wish to remain British. Britain's reoccupation of the islands, under Operation Corporate, was simply defending sovereign territory.

The Argentinians overran the Islands and installed a garrison of more than 12,000 men. Britain despatched an amphibious force over 7,000 miles. Just one small island (Ascencion) could be used as a staging post. The landing force initially consisted of just four infantry battalions, one field artillery battalion, and virtually no armour. At most there were seven infantry and two artillery battalions.

The British air component initially consisted of 28 Royal Navy Sea Harriers, later reinforced by 6 RAF Harriers. Argentina lost about a hundred aircraft. 11 were destroyed on the ground in a Special Forces raid. 29 were captured when the Argentinians surrendered. 30 were shot down by Harriers, and 17 by SAMs from warships. Britain lost 10 Harriers and 12 helicopters. Five of the Harriers, including three from the RAF, were shot down. All of them were shot down by ground-based air defences. (Between 1945 and 2000 the RAF shot down a total 10 enemy aircraft. 12 of its own aircraft were shot down. No RAF aircraft has shot down an enemy aircraft since 1948.)

The RAF mounted a long-range bombing operation using V-Bombers operating from Britain. Its purported mission was to deny the use of the one long airfield on the Islands, at Stanley, to Argentinian jets. Those aircraft had already been withdrawn, due to the threat of shelling by British warships. The real objective of the bombing seems to have been to show British politicians that the RAF was still relevant. Political signalling, once again. It was greeted with indifference, disdain or even derision.

British submarines soon identified the Argentinian 8-inch gun cruiser ARA *Belgrano* due to the acoustic signature of the tanker accompanying it. *Belgrano* was sunk by HM Submarine *Conqueror*. As a result the Argentine Navy withdrew its surface units, including its one aircraft carrier. The Royal Navy could then operate its two carriers east of the Falklands, and frigates and destroyers west of the Falklands, relatively safely. Nevertheless four British warships were sunk, and six more damaged. So were three other vessels. Argentine aircraft used a mixture of bombs and air-launched guided missiles. Unfortunately all but one of the three Chinook heavy lift helicopters were lost on the MV *Atlantic Conveyor*.

The British landing operation went fairly smoothly. One (paratroop) battalion cleared a position to the south, at Goose Green. It, and the remainder of the force, then had to march east across 50 miles of rough terrain, in atrocious weather, before attacking Argentine forces dug in on hilltops around Stanley. Five battalion attacks were conducted, over three nights. Each position was defended by an enemy battalion. All five attacks were completely successful. The Argentinians then surrendered.

The six attacks can be analysed numerically. The attackers lost between nine and 63 killed and wounded. The defenders lost between 80 and 160. There was no correlation between defenders' and attackers' casualties; nor between defenders' casualties and the duration of the battle. There *was* an inverse relationship between the speed of the attack and attackers' casualties. That is: the faster the attack, the lower the attackers' casualties. The correlation was particularly strong on the one occasion where the attackers achieved surprise. That was 42 Commando Royal Marines' attack on Mount Harriet. British casualties there were two dead and seven wounded.

Overall, the British lost 255 dead and 777 wounded. The Argentines lost 649 killed (323 on the *Belgrano*), 1,657 wounded and 11,313 prisoners of war. Success was due to genuinely global strategic reach coupled to professionalism. At practically every level from rifleman upwards, the British were simply better than their opponents.

In 1990 Saddam Hussein invaded Kuwait as a consequence of the Iran-Iraq war. Iraq was short of money and had a legitimate grievance with Kuwait over oil revenues. Nevertheless, the invasion was a clear violation of sovereignty. The UN voted almost unanimously to support military action. The USA assembled a coalition force to liberate Kuwait in what became the First Gulf War.

The Iraqi Army deployed 43 divisions. 24 were in Kuwait, 10 were along the Saudi-Iraqi border and nine were in reserve around Basrah. Coalition ground forces included seven US Army, two US Marine, one French, one British, and about five Arab divisions (for a total of 16).

Around 2,250 Coalition aircraft, of which 1,800 were American, faced about 550 operational Iraqi aircraft. The 39-day air campaign gained air superiority almost immediately. 535 so-called 'strategic' targets were attacked, but they included power stations, transport infrastructure and oil installations. Very few Coalition aircraft were lost. Iraqi SAMs fired about 1,800 missiles per aircraft shot down.

By the beginning of the ground campaign the Iraqi army had lost about 140-160,000 men (of about 360,000). The great majority simply deserted and went home. By the end of the campaign Iraq had lost 76 per cent of its tanks. However at that point the picture becomes unclear. In one sample 10-20 per cent of Iraqi AFVs had been hit by air attack. 50 per cent were not hit at all, and 30-40 per cent had been hit by ground fire. In the Iraqi withdrawal from Kuwait City over 1,400 vehicles were damaged by air attack. Only 14 were tanks and 14 were other armoured vehicles.

Coalition air attacks did *not*, however prevent Iraqi armoured formations manoeuvring. They did not cause decisive attrition to Iraqi mobile reserves. They did not demoralise those reserves, critically degrade Iraqi command facilities, nor reduce the combat power of those reserves.[3]

It is clear that massive one-sided numerical and technical aerial superiority did have a major effect on the campaign. Coalition air forces had benefitted from a once-in-a-generation uplift in technical and tactical capability. They also had a massive numerical superiority that reflected the end of the Cold War. The severe losses inflicted on Iraqi forces withdrawing from Kuwait City was an excellent example of air attacks devastating a fleeing enemy. However it is not true that air attack had a decisive effect. It did not resolve nor settle the issue; tactically nor operationally. Indeed one airpower theorist wrote, absolutely clearly, that '*ground* manoeuvre was decisively successful'[4] (emphasis added).

The subsequent Gulf War Airpower Survey (GWAPS) was mandated by Congress. The analysts did a very good, thorough and dispassionate job. US Air Force officers then tried to massage it to demonstrate the effectiveness of air attack. GWAPS contains several throwaway lines like 'x *must have* had such and such an effect' (emphasis added) or 'common sense dictates that ...'. Where facts alone do not support conclusions,

3 Daryl G Press, 'The Myth of Air Power in the Persian Gulf War and the Future of Warfare'. *International Security* Vol 26 No 2, 10-12 and 26-38.

4 Gray, *Airpower for Strategic Effect*, 212.

expressions like that are a clear appeal to emotion rather than logic. There is evidence both of tampering with the analysts' findings and of trying to bury the report.

We also see airpower theorists conflating activity with outcome. We see that so many aircraft, so many sorties and so many tons of ordnance 'decided the fate of the battle well before the ground offensive began.'[5] Well, they didn't, and that is an excellent example of a classic fallacy. Activity is not outcome. The effectiveness of air attack in Operation Desert Storm is a long-standing, well-embroidered myth.

The main effort of the Coalition ground attack was a brilliantly conceived envelopment by the nine US Army, French and British divisions. That was a clear example of the effect of large armoured forces conducting joint, theatre-level ground-air operations. It was also a very good example of the qualitative superiority of American and Coalition soldiers, doctrine, training and equipment. It was decisive. It did resolve, or settle, the war. Writ large, the war was also a clear demonstration that America had buried the demons of Vietnam.

However, analysis suggests that most of the attacking divisions were inefficiently large. Furthermore, the corps structure was ponderous. For example, the US VII Corps undertook two simultaneous operations with two divisions each, and a third operation with its fifth division. Its armoured cavalry regiment did something else. It seems that more, smaller divisions in more, smaller corps would have been more effective.

High-level air-land coordination was flawed. There was little coordination between the initial air offensive and the land operation. During the ground offensive there was an argument between the land and air components over the location of the Fire Support Coordination Line. The argument probably had a negative impact on the overall outcome. Although apparently technical in nature, the disagreement seems to have largely been a matter of interservice rivalry.

The First Gulf War heralded what was seen, cautiously, as a potential 'brave new world'. The yoke of Communism just had been lifted. Liberty and democracy improved around the globe. Economic growth and improved literacy levels followed. There were also understandable pressures to reduce defence spending and enjoy a 'peace dividend'. In

---

5 John Andreas Olsen in Gray, *Airpower for Strategic Effect*, 212, footnote 47.

Britain, for example, the defence budget fell from about five per cent of GDP to about two and a half percent. The Armed Forces shrank by perhaps a third.

There were still trouble spots around the world. There was also a feeling that those hotspots should be policed. The demise of the Soviet Union made it easier for the West to do so. Numerous 'peace support' operations were initiated. NATO developed a peace support doctrine. Although praiseworthy, it had some negative consequences. Looking outside our period briefly, in late 2003 and 2004 western nations operating in Iraq mis-identified the early stages of an insurgency as a 'fragile peace', or similar.

America's involvement in Somalia in 1993 demonstrated that its armed forces were not yet well-prepared for 'the savage wars of peace'. The term had been coined by the British poet Rudyard Kipling about a century earlier. Ironically Kipling was referring primarily to the American colonisation of the Philippines.

In Yugoslavia, Slobodan Milošević manipulated electoral law to take personal control of much of what was officially a federal system. Yugoslavia descended into a multi-party civil war. It displayed much of the viciousness and bestiality that civil wars often do. Two aspects merit discussion.

The first is the effectiveness of international humanitarian law. In the former Yugoslavia and (for example) Rwanda there were indisputable instances of what we now describe as crimes against humanity. Yet it took years, sometimes decades, to bring those responsible to account. So, from the perspective of a local politician with grandiose ambitions, what real sanction is there against mass murder, genocide or similar atrocities? In practice there is some chance that, years from now, you *might* be indicted in some faraway court and you *might* spend a long time in prison. Would-be despots and dictators might, rationally, accept that risk.

The second issue is the continuing institutional arrogance of air forces. 'Arrogance' means 'having an exaggerated opinion of one's own importance or abilities'.[6] Here we focus on 'abilities'. In the Balkans NATO air forces played a role in bringing parties to sign the Dayton Accords in 1995 (under Operation Deliberate Force). So, however, did cruise missile

6 OED.

strikes from a US warship. So did deploying NATO artillery units on Mount Igman above Sarajevo.

In 1999, Milošević moved against ethnic Albanians in Kosovo. NATO air forces bombed Yugoslav (meaning largely Serbian) armed forces and infrastructure targets, including some in Belgrade. The NATO air commander, General Mike Short, seemed to think that the conflict could be resolved entirely by airpower. The bombing continued for about 11 weeks. Once again, the air forces ran out of targets. Intense diplomacy, linked to the threat of a ground invasion, persuaded Milošević to desist. He was presented with political conditions which he could accept. He did, and backed down. Air attack *had* had an effect. It was political signalling. It demonstrated resolve; nothing more.

On reflection, it was ridiculous. After several weeks' effort the most powerful nation on earth, together with several of its rich allies, persuaded the 89th richest nation to back down. A triumph of airpower? Big deal. That is triumphing over the successful ascent of a molehill. NATO could, largely, control the damage its air forces did to Serbia. What it could not control was Serbia's political response.[7] Air attack is not like that. Using air raids for coercion is generally flawed. Airmen don't see that.

What observations or deductions can be drawn about war and warfare in the last 25 years of the 20th century?

The first is the importance of political settlement. Northern Ireland and Kosovo were resolved almost entirely by political negotiation. Collective armed violence played some part in shaping the discussion and the outcome. There, and elsewhere, the violence had a social and political legacy. However, when one stands back and looks at the eventual settlements, one largely sees the results of political discussion; not those of the use of force.

But what does one do when the other party will not negotiate? Force (collective armed violence) can play a role. Britain liberated the Falklands (that is, reoccupied them by force). Argentina was then powerless to re-invade. The American-led UN Coalition liberated Kuwait. Iraqi armed forces were damaged so much that Saddam could no longer undertake military operations against his neighbours. In such cases, to be blunt, the

---

7 Gray, *Airpower for Strategic Effect*, 217.

subsequent political settlement does not matter much. (Who remembers whether Argentina agreed to anything? Or Saddam?).

However, wars do not resolve issues for all time. 12 years later an American coalition attacked Iraq again. If there is a lasting peace in Northern Ireland, or Kosovo, it is more due to the political agreements that were reached (and the economic, social and cultural developments which followed) than the application of violence. Indeed, in such cases the use of violence typically generates a legacy that is bitter rather than positive. That is a good reason for restraint in the use of force. The Serbs, for example, generally do not relish having been bombed in 1999.

In applying collective armed force – violence – for political purposes, professionalism makes a difference. At one level, professional armed forces tend to beat conscripts. They do so at relatively little cost (look at the casualty figures for the Gulf War or the Falklands, for example). At another level, well-equipped and well-trained forces tend to win, and win quickly. Most regular (as opposed to irregular) wars fought since 1945 were won within 90 days. They were won by the better-trained and better equipped force. Not, that is, by the larger force. Professional armed forces also seem to show more restraint in the use of force. That leaves less of a legacy of resentment.

At another level – the strategic level – professional armed forces tend to give better advice to their governments, so the resulting strategy is better as well. That statement, admittedly, would benefit from further exploration. It risks circular argumentation. Did Saddam's generals advise him badly because they were not professional? Or were they not professional because they advised him badly? Or is the real issue that he did not listen to their advice?

Unfortunately professionalism has continued to persuade air force commanders that air attack can win wars. That is, that it can have decisive strategic impact. It can't. It hasn't. That was demonstrated in Iraq and Yugoslavia. In other conflicts air forces either didn't have sufficient reach (such as in the Falklands), or were largely irrelevant (Northern Ireland). Sadly that didn't stop air force commanders thinking that air attack can win a war; or that it has; *or maybe that it will next time*. In 2006 the Israeli Air Force tried, and failed, to win in Lebanon by bombing alone. Some things haven't changed.

The sadness is that, when integrated into maritime-air or land-air operations, air forces have sometimes had a devastating impact. Slessor broadly had it right in 1936. We will examine the institutional dynamics of air force doctrine in Chapter 13.

11

CONVENTIONAL LAND WARFARE

This book is not, in the first instance, a work of history. It is an analysis. To repeat: what can we learn from war and warfare in the twentieth century? Having completed a chronological review, we now consider the question by domain. Chapter 11 looks at conventional land warfare. After some initial observations it looks at empirical data. That is, what do the numbers tell us? We then look at mechanisms. How can we characterise the processes by which land forces fought: on the battlefield, and across campaigns? Finally, we draw some observations and deductions. We shall not consider the tactic of raiding here. That will be discussed in chapters 12, 13 and 15.

At first sight, three major issues stand out. Firstly, and at the lowest tactical levels, the rifled bullet still dominates the battlefield. Small arms fire (and the threat of it) still forces soldiers to disperse, conceal themselves and move cautiously. Dismounted combat, of itself, remains slow and generally indecisive. Developing tools and mechanisms to overcome that was a major (if generally implicit) problem for land warfare throughout the twentieth century.

The second issue, however, is almost the converse. Armoured warfare works: especially at the campaign level. Writing well into the twenty-first century (in 2024), we can say that since 1945 only four conventional wars have lasted more than 90 days. Largely through the use of armoured formations in integrated land-air operations, conventional wars have generally been quick and decisive.

The third issue is a reflection on the previous two remarks. At its simplest, war is a mixture of violence (the use of weapons) and movement. Weapons usage, however, is tactical. No weapon, including nuclear weapons, has significant direct effect more than a couple of kilometres from its point of impact. The great majority have no effect beyond a few metres. Damage caused by weapons usage is tactical. Calculating how to benefit from that is a key issue for operational and strategic planners. Conversely, however, *movement* may be tactical, operational or strategic.

Thus at this stage we can make a tentative initial observation. Conventional land warfare is largely an issue of violence and movement. The movement, however, seems to have the most significant effect.

Napoleon said that 'the strength of an army, like the quantity of movements in mechanics, is estimated by the mass multiplied by the quickness', or similar.[1] He meant that speed of movement is just as important as combat power; if 'combat power' is measured by things like numbers, calibres of weapons, and so on. The Soviets considered that a force that decides and acts twice as fast can defeat one that is five times its size.[2] That statistic largely explains how Red Army units and formations were regularly beaten by smaller German forces in the Second World War. Those two observations (Napoleon's, and the Red Army's) point to an emphasis on speed of decision, action and movement long suspected, but never quite captured, by military thinkers.

We now step away from historical research to numerical analysis. We move from the traditional methods of military history towards statistics. A century gives us a lot of data from which to gather statistical insight. We do so cautiously. We rely on respectable, well-funded research. Historians were employed to gather empirical data from historical records. Mathematicians then drew statistical inferences, which were then checked by the historians for relevance. The results can be surprising. There are two main western sources of this 'historical analysis'. The first comes from the work of the British David Rowland and his colleagues.[3, 4, 5]

Amongst other questions, Rowland asked which factors were correlated with overall success at the theatre and campaign level. The

---

1 C T Montholon, *Memoirs of the History of France During the Reign of Napoleon*. (London: Colborn, 1824), IV, 290; in War in History Vol 30, I, 30.

2 G Hyde, Notes from a briefing given to the British General Staff, Moscow 1990. In an article by Major General John Kiszely, Journal of the Royal United Services Institute, Vol 144 No 6, December 1994.

3 D Rowland, L R Speight, and M C Keys, *Manoeuvre Warfare: Conditions for Success at the Operational Level*. Unpublished and undated, circa 1993, 5.

4 See also L R Speight, D Rowland, D, and M C Keys, *Manoeuvre Warfare: Force Balance in Relation to Other Factors and to Operational Success*. J Military Operations Research Vol 3 No 3 1997, 31-46.

5 I worked with David Rowland, and some of the other authors of such work, between 1999 and 2001.

analysis looked at 159 land campaigns as far back as the American Civil War. The results were surprising. Force ratios, better intelligence, artillery superiority and so on all had some effect. However what drove the results were combinations of shock, surprise, opportunistic exploitation and control of the air. As an example: all other things being equal, a favourable force ratio might make perhaps a ten per cent difference in the probability of campaign success. However the combination of surprise, air superiority and opportunistic exploitation led to a 98 per cent probability of campaign success. Put simply: we can see what wins. Shock was another significant factor. It is closely related to surprise. In the attack, shock led to a roughly 65 per cent further reduction in the effectiveness of the defence.

'Surprise' relates to the unexpected. In this analysis, 'unexpected' refers to means (weapons); methods (tactics); direction of attack (particularly from the flank or rear); and timing (particularly if early). Surprise was the single most important factor. It could have the same effect as a force ratio of 2,000 to one.[6] Where surprise occurred, the probability of success was largely independent of force ratio. That suggests that surprise can render force ratio largely irrelevant. However, see below.

'Shock' has two aspects. Shock *action* is the sudden, concentrated application of violence. It is typically the neutralising effect of concentrated indirect fire. Alternatively it could be the stunning, frightening effect of a massed charge. The charge might be by horsed cavalry, tanks, or (occasionally) infantry.

Shock *effect* is observed where the target is numb, unresponsive or perhaps behaving irrationally. Shock affects individuals. Critically, analysis showed that it can also affect units, and formations. The chain of command is paralysed. It is deluged by reports of loss and failure. Commanders and staff may panic. Even if not, they take time to process events. During that time the command lacks direction. Shock effect is best created by some combination of surprise and shock action. In some circumstances such combinations can have the same effect as a force ratio of 1,200 to one.

6 R C Blues, D Rowland, M C Keys and R C Dixon, *Historical Analysis Paper for PPSG Case Conference on ISTAR*. Centre for Defence Analysis document CDA/HLS 44/393 dated November 1995, 6.

Such statistics have limited value. Force ratios of more than about six to one are very rare. In effect the numbers tell us that shock and surprise are greatly more effective than anything else that occurs on the battlefield.

Shock and surprise are, of themselves, transient. They wear off. They must be exploited to generate concrete gain before they do wear off. This is the substance of 'opportunistic exploitation', referred to above, which we will consider later in this chapter. We will look at the relevance of air superiority in Chapter 13.

A different view of this issue comes from the work of American historian Archer Jones.[7] His review of 4,000 years of military history suggests very strongly that without a mechanism for shock action, combat tends to be protracted and indecisive. We saw earlier that by 1914 the dominance of the rifled bullet meant that the largely infantry and cavalry armies of the period generally lacked the ability to inflict shock.

In a different area of research, Rowland and his colleagues looked at the actual effectiveness of weapons on the battlefield, as opposed to during trials.[8] As an example, we can count how many hits soldiers achieve on a firing range. Trials and historical analysis show that their performance on a field firing area is roughly one tenth of what it is on a range. If the soldiers are fired at, their performance is reduced by another factor of ten. If the enemy has machineguns, there is a further reduction by a factor of ten. There is another tenfold reduction if enemy tanks are firing at them.

Clearly, in some conditions infantrymen are extremely unlikely to hit anything. To that, add the fact that in a typical battle many infantry soldiers do not actually see an enemy. That may simply be because relatively few are in the front line at any stage.

Historical analysis also pointed out that in close (typically wooded or urban) terrain the defenders were at a particular disadvantage. That is partly because their fields of view are generally reduced. The same is true for the attackers. On balance, however, this favours the attackers. The defenders' ability to engage at long range, and for relatively long periods, is much reduced. Furthermore, the attackers are typically more likely to be able to surprise the defenders.

---

7 Archer Jones, *The Art of War in the Western World* (London: Harrap Ltd, 1988), passim.

8 D Rowland, 'The Use of Historical Data in the Assessment of Combat Degradation'. *Journal of the Operational Research Society*, 38.2, 1987.

Quite separately, the American Colonel Trevor Dupuy and his colleagues conducted numerical analysis which was largely based at the formation level.[9] Dupuy's findings are generally consistent with Rowland's. Any differences are typically due to the two groups asking slightly different questions.

Dupuy found that the outputs of battles and engagements are not driven by force ratio. '[A] direct cause-and-effect relationship between force ratio and [casualty] exchange ratio does not seem to exist.'[10] Loss exchange ratios are a result of battle outcomes, rather than force ratio or some putative 'combat effectiveness'. Put simply: if you win a battle, your losses will tend to be lower (and the opponents' higher) than if you lose.

Dupuy found that, at divisional level, it was very rare for western attackers to win at a force ratio below one to one. It was unusual to lose above a force ratio of two to one. By comparison, in the Second World War the Soviets found it difficult to win without a considerably larger numerical advantage. The Germans would succeed against the Soviets at force ratios appreciably lower than they needed in the west.[11] Thus force ratios do have an effect. However, as we have seen, shock and surprise had an effect much larger than any likely force ratio.

Surprise contributes to success. Dupuy found that the attackers won in 50 per cent of the battles he considered, if they did not achieve surprise. They won 86 per cent of the time if they did. He considered that surprise was relatively rare. At the theatre level surprise occurred in about ten per cent of the campaigns he considered. However it occurred in more than 20 per cent of divisional engagements, almost 30 per cent of battalion engagements, and 60 per cent of lower-level engagements. The latter was, admittedly, a very small sample in Dupuy's work. This suggests, however, that surprise is not just transient but also, typically, a localised phenomenon.

There were differences between armies. In Italy in the Second World War, American forces surprised German forces 11 per cent of the time: British forces surprised Germans only six per cent. However American forces were surprised *by* German forces 45 per cent of the time. British

---

9 Christopher A Lawrence. *War by Numbers. Understanding Conventional Combat.* (Lincoln: Potomac Press, 2017).

10 Ibid, 78.

11 Ibid, 11.

forces were surprised by Germans in 13 per cent of engagements. Not least, the Germans were clearly better at achieving surprise.

Some of Dupuy's findings were negative, but nonetheless informative. Information advantage is relatively unimportant. It does not necessarily favour the attacker over the defender, nor the converse. It makes whoever holds the advantage more likely to achieve his mission in a small number of cases. It does not lead to higher rates of advance. However it doubles the chance of achieving surprise. In turn surprise leads to a doubling of the rate of opposed advance. As we have seen, it also leads to a higher chance of winning. So, technophiles should be cautious. Information advantage, perhaps gained by digital electronics, is not a winner of itself. It is a winner largely because of what it can do: allow the generation of surprise.

Dupuy also analysed urban combat. He found that it is *not* more intense than combat in rural terrain. Tank losses are *not* higher. The rate of advance *is* slower in urban terrain. However, across a campaign that effect is lost. Urban areas are typically enveloped, then abandoned by the defenders (who would otherwise risk encirclement). As Rowland also found, attackers' losses are reduced. Defenders' losses rise.

At this stage we really must comment. Many historians are simply wrong. They typically describe urban combat as being particularly lethal, intense, and a graveyard for the attackers. Well, some urban battles might have been like that, but that is not the overall picture. It is not typical. In every case which we have discussed, the attackers' casualties were lower than the defenders'. In Rowland's work, that was *every* case for which he and his colleagues could find accurate data. In general, historians have done a poor job in this area. They may have served us poorly in other areas as well.

We now turn from global, statistical treatment to issues of battlefield mechanism. We start at low level, with an attack by an infantry section (or American 'squad'). The section has a light machinegun. To suppress a couple of enemy in a trench 600 metres away the gunner needs to hit an area about 1.3 metres wide and 1.1 metres high about seven times out of ten, firing bursts roughly every nine or ten times a minute. That equates to an area just 1.3mm by 1.1mm at arm's length.[12] The gunner should fire

12 Nominally 60 cm from the eye.

about 30 rounds per minute. (As an aside: if you are an infantry officer, can your machine gunners do that?)

Whilst the machine gunner does that, the rest of the section should infiltrate forward, using cover, until close enough to throw in grenades. They should then rush in and kill or incapacitate any survivors with small arms and bayonets.

That scenario might sound unlikely, but it is typical of training methods used by western armies in the 1950s through to the 1990s. The riflemen may have to use rifle fire to help them cross open ground. Machine gunners may fire many more rounds. The evidence suggests that if they do, their fire will be less accurate and therefore *less* effective at supressing the enemy. Instead of using a dedicated machinegun team, the section may employ half-section 'fireteam' tactics. Evidence suggests that their suppressive fire is, in practice, more likely to break down at some stage. The attackers will take more casualties.

Go up the scale. A platoon attacks an enemy section. One attacking section suppresses the defenders. The remainder move round the flank to assault the enemy. The enemy might withdraw before that happens, but (with thought) the attackers can engage them as they do.

Such an attack may well not work. One attacking section, possibly in the open, is trying to suppress a dug-in section of defenders. That probably won't work by itself. Bombs from a light mortar, dropping into the enemy's trenches, might make the difference. Heavier mortars, or artillery fire, probably would. In general, frontal attacks without external fire support are not likely to succeed.

Writing after the Second World War, former Wehrmacht officers stressed repeatedly that infantry units cannot attack successfully without fire support.[13] Indirect fire support (from mortars or artillery) of over 75mm or so calibre will suppress. If over 100mm, it will probably neutralise as well. The problem is to coordinate the fire with the movement of the infantry, in real time, so that the attackers are never caught in the open. The solution to that lies in having enough indirect fire controllers, and having them in the right place at the right time. At higher levels the defenders' artillery needs to be suppressed, by counterbattery fire, so that it cannot

13 Eike Middeldorf. *Handbuch der Taktik. Fur Führer und Unterführer.* (Berlin: E.S. Mittler & Sohn, 1957). Author's translation. 93.

engage the attacking infantry. Counterbattery fire was typically undertaken by medium and heavy artillery.

Tanks escorting the infantry, firing HE, destroy (or at least neutralise) enemy trenches, bunkers and occupied houses. That speeds the process up considerably, and reduces attackers' casualties.[14] The combination can induce shock effect.

Consider a company attacking with four sections in the first echelon (Figure 11-1).

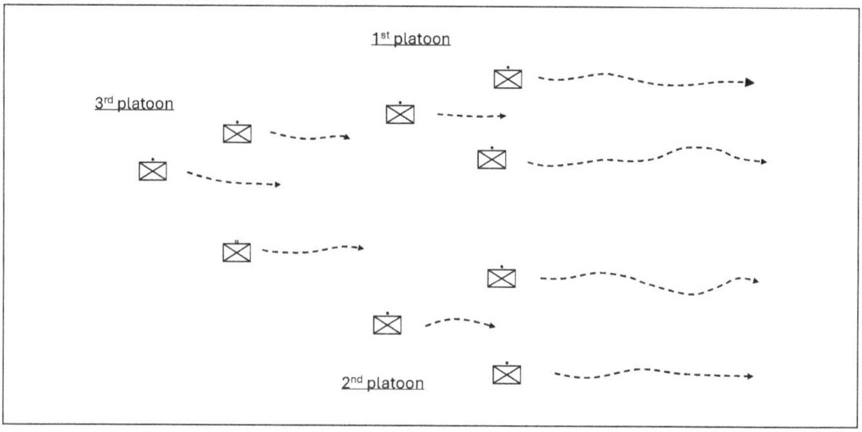

Figure 11-1. A Company Attacking with Two Platoons and Four Sections Leading.

The company commander does not allow his platoons to advance past any enemy position still resisting. The attack then slows to the pace of the slowest section. That is stupid.

Alternatively, the company commander encourages platoons to bypass wherever possible. Sections outflank surviving enemy and engage or attack them from the rear. Every such action speeds up the progress of the attack. Additionally, some sections may find that there is no-one in front of them. They penetrate rapidly into the enemy position, creating surprise, and causing alarm which translates into panic: shock. That is infiltration. It is probably the most important lesson of the Great War.

The main risk is to the flank of the penetrating element. The counter to that is support from the next element in the column: See Figure 11-2.

14 Rowland, *Combat Degradation*.

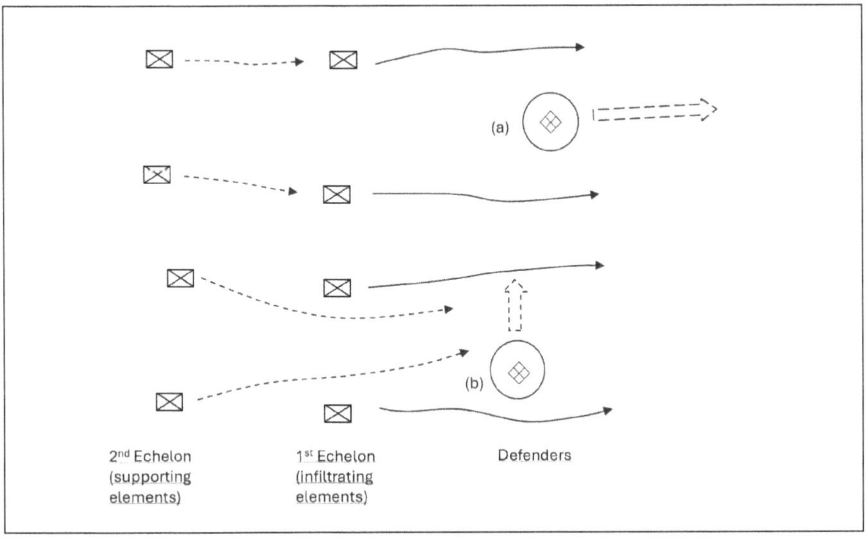

Figure 11-2. Column Tactics

Notes:

a. This defending element resists until bypassed but then withdraws, under risk of being attacked in the rear.
b. This defending element attempts to counterattack a bypassing attacker but exposes himself to attack from flank and rear.

We begin to see the elements of Liddell Hart's 'expanding torrent'.[15] It can work very well at such low tactical levels. Surprise is a battle winner and *the* campaign winner. We begin to see how. Opportunistic exploitation from the lowest levels – well below company – leads to ever-greater opportunity. It can also lead to surprise at ever-higher levels.

This account has centred on the infantry. It reflects the dominance of the rifled bullet. In some circumstances better alternatives are available. An armoured advance led by tanks may well be much faster and result in greater shock and surprise. Infantry cooperating closely with the tanks can suppress antitank weapons and winkle out pockets of opposition. The tanks can lead the infantry to the defenders' flanks and rear. It can work well. It can work fast. It takes a lot of training. Ideally it takes tanks, infantry, sappers, gunners and others who are used to working *with each other*.

15 Captain B H Liddell Hart, 'The "Man-in-the-Dark" Theory of Infantry Tactics and the "Expanding Torrent" System of Attack,' *Royal United Services Institution. Journal*, 66:461, 1-22.

Even the most successful attack of this kind is only a local success. What happens before and after that?

The attackers may have to approach from considerable distances: perhaps hundreds of kilometres. To do so, they will need drills for long marches (that is, vehicular road moves.) They should be linked to drills for deployment from the march, and drills to transition from the attack into exploitation and pursuit. (The British had, and have, neither.)

Scouts should lead. To repeat: surprise is a battle winner and *the* campaign winner. Reconnoitring by stealth allows the identification of opportunities for surprise. That means advancing where the enemy is absent. It means attacking the enemy's flanks and rear, thereby exploiting surprise. Finding where the enemy are *not* is more important than finding where they *are*. That may allow bypassing and moving rapidly into the enemy's rear. An advance guard should be used to clear minor opposition without delaying the advance of the main body. It should infiltrate rapidly in order to do so.

Indirect fire controllers should accompany the advancing force. Guns and mortars should move forward to keep in range. Sappers should clear obstacles, although leading elements should bypass such obstacles where possible. The logistic chain should extend.

An attack is only decisive if it breaks through an enemy position and allows exploitation. Repeated attacks on prepared positions are rarely decisive in themselves. Think 'Normandy'. Repeated British and Canadian attacks allowed the US Army to break out. The subsequent American exploitation decided the Normandy campaign. At some stage an exploitation may become a pursuit. Units and formations should then be given distant objectives which have some operational relevance. A theatre-level encirclement would be an example.

We now turn to defensive operations. Western armies have been able to break through single, linear defensive systems quite easily since 1915. (The main problem was to turn a break *in* to a deep defensive system into a break*out*.[16]) Several measures were developed to prevent or mitigate such a breakthrough.

A screen force observes and reports the enemy's movements. It provides intelligence and protects against surprise. Conversely a guard

16 *Report of the Committee on the Lessons of the Great War* (The Kirke Report). The War Office, 1932, 11.

force has a different (but related) function. It defeats enemy reconnaissance. That further prevents the enemy achieving surprise. The guard force delays the enemy advance. That may gain time: for example, to complete withdrawals or prepare defensive positions. The guard force may set the enemy up for a counterattack by withdrawing in a designated direction.

Most importantly, a guard force prevents the enemy locating the main defensive position. We shall see the relevance of that later. Indirect fire and air attack delay the attackers and cause attrition. Sappers create obstacles and demolitions, which cause further delay. They may focus the direction of the attackers' advance, although there is little evidence that they do. Local counterattacks cause delay and attrition, and impose caution. They should have limited objectives, not least because the guard force will be needed to act as a reserve later in the battle.

It is clear that covering force operations (the actions of screen and guard forces) have an important function. They are unlikely to be decisive but they may form part of a decisive defence, in which the attacker is definitively halted. If the attack is halted the attackers cannot achieve (or indeed reach) their immediate objectives.

A defence has to defend something, somewhere. That is best defined by a main line of resistance (MLR). At its most simple the MLR is a more-or-less continuous chain of infantry section positions. Its strength is based not on the weight of fire it produces, but the interlock of the fire of its section machineguns. Each machine gun protects its neighbouring section's front. That exploits and maximises the effectiveness of the rifled bullet. Riflemen protect the machineguns. They employ shoulder-fired light antitank weapons (LAW) to avoid being overrun by AFVs. If the chain has gaps it can, and will, be penetrated or bypassed.

The defenders also need heavier antitank weapons to stop armoured attacks. That is the function of a network of antitank guns (or, more recently, guided missile launchers) along or around the MLR. The antitank weapons are in practice the backbone of the defence.

Almost any defensive position, once located, can be neutralised and penetrated. To avoid that it must be concealed. That means preventing reconnaissance of the main defensive position. Additionally, outposts forward of the MLR conceal, deceive and mislead. They provide warning. They inflict casualties. They do not fight to the last man. Western armies seem to have forgotten about them.

Any chain can be broken. Platoon, company and battalion reserves counterattack to restore the MLR. They do so immediately and from unexpected directions. Tank support is useful, but no immediate counterattack should be cancelled due to shortage of infantry. Immediate dismounted counterattacks have an effect beyond mere numbers.[17] They allow defenders to surprise and attack the attackers.

More deliberate counterattacks are planned to achieve definite objectives. Is a given terrain feature critical to the functioning of the defence? If so, it should be counterattacked if lost. If not, thinning out a ruptured MLR can be used to form a new outpost line, forward of a new MLR established in depth. The new outpost line then conceals, misleads and deceives in relation to the new MLR.

Multiple defence lines arranged in depth can delay the attacker and cause attrition. However in almost all cases a defence based on counterattacks is stronger than one based solely on depth. In a deep defensive position based on the counterattack, the rearward positions provide cover for the counterattacking troops. They also form the basis for a protracted defense if necessary.

Conventional combat is obviously unutterably complex. Every rifleman, every machinegun, fire controller, tank and antitank weapon plays a part. They are numerous. They all interact. That complexity cannot be understood, let alone planned for, in advance. Yet attacks can succeed. Defences can succeed. How?

The key lies in the recognition and exploitation of low-level opportunity. For the attacker, a route to an open flank or an unprotected rear may be narrow. Where one man can go, however, a section can follow. Where a section can go, damage can be done; and exploited. The route may only be open for a brief moment. That moment, that opportunity, should be seized. That will create a situation which is unexpected. It should be exploited in order to achieve some part of the higher commander's intentions. 'Exploitation' is purposeful activity to further the overall commander's objectives. For the defender, immediate counterattacks surprise and shock (hence halt) the attacker and prevent enemy exploitation.

Opportunistic exploitation relies on initiative. In this sense, 'initiative' is taking unforeseen (and potentially unauthorised) action to further the

---

17 Eike Middeldorf, *Taktik im Russlandfeldzug. Erfahrungen und Folgerungen.* (Berlin: E.S. Mittler & Sohn, 1957). Author's translation. 148.

overall commander's objectives. Initiative requires a permissive, and supportive, culture with regard to authority. That should not be taken for granted. In some armed forces it will simply not occur.

Initiative and exploitation should create surprise. To repeat once more: surprise is a battle winner and *the* campaign winner. Surprise can be seen as an 'attentional blink'. Commanders and staffs pause, perhaps subconsciously, to identify the new situation. They must then decide how best to deal with it, and turn that decision into action. That takes time.

Surprise at low level can and should, where possible, lead to surprise at higher levels. Some commanders simply cannot cope with surprise. Their expectations and their sense of self will be violated. They may react irrationally. They may blame others, thus making the situation worse. Some commanders may panic. Other commanders may be more phlegmatic. Can anyone really imagine a Rommel or a Patton being perturbed by anything the enemy might do? Rommels and Pattons, however, are rare.

We observed that, at its most basic, combat is some combination of violence and movement. Formation-level armoured warfare works. It can win campaigns quickly. That is despite the primacy of the rifled bullet. It does not work simply because tanks are bulletproof. Antitank weapons, crewed by dismounted soldiers, can stop tanks. They have generally been more effective than tanks in destroying enemy tanks. Yet, overall, the twentieth century shows us that some armies developed armoured forces (and tactics) which can survive, fight, move and win campaigns quickly. Add to that 'relatively cheaply'; certainly in relation to the carnage of the Great War. Their successes depend on three requirements.

The first requirement is a well-trained all-arms force. Sappers, gunners, maintainers, suppliers, medics and the rest all have critical parts to play. They need to know their parts, and how to play them in the all-arms team. That means frequent all-arms training. You cannot assume that any team will work well together unless it trains together. It should train again whenever new members join the team.

The second requirement is doctrine for all aspects of the process. Some elements may be missing. The British, to repeat, have no drills for deployment; nor for transitioning from attack to exploitation and pursuit. Most armies seem to have forgotten about outposts.

The third requirement, however, is a flexible command culture. It permits, indeed expects, opportunistic exploitation. It rewards initiative. It tolerates well-intentioned mistake. Does your army do that?

What are the main observations and deductions that we can draw from conventional land warfare in the twentieth century?

Despite many advances, dismounted combat remains slow and indecisive. That is largely due to the dominance of the rifled bullet. As a result, infantry units are largely unable to attack without support. The key to that support is HE: in the right place; at the right time; and for long enough. That fire support may be direct or indirect. Nevertheless somebody, somehow has to suppress or neutralise enemy small arms fire so that the infantry can get forward. If not, dismounted combat will remain slow and indecisive.

Armoured formations, as the ground element of a land-air force, can win campaigns quickly and decisively. As described above, the three requirements for that to work are: well-trained all-arms formations; well-written, comprehensive doctrine; and a flexible command culture. Whether winning campaigns means winning the war is a question of strategy. It is not primarily an issue of conventional land warfare.

Fighting is a combination of violence and movement. Human phenomena like shock and surprise do, and should, play a major part in its conduct and its outcome. However violence and damage are tactical. Movement may be tactical, operational, or even strategic. The major deduction from this chapter is that, at the tactical and operational levels, fighting is most effective when the emphasis is on movement rather than violence. Some fighting is probably inevitable, but the emphasis should be on manoeuvre: movement to a position of advantage. That means movement to create an *operational*-level advantage, whenever possible.

Simplistically: in the early stages of the Great War armies could not achieve shock and surprise on the battlefield. They learnt to do so, using artillery and assisted by tanks. What they could not normally achieve was operational success through manoeuvre. There were exceptions, notably Romania in 1916 and Palestine in 1918. In the Second World War some armies could achieve shock and surprise on the battlefield, largely by using armoured formations. Critically, some could exploit that up to the theatre and campaign level. That has been repeated several times since 1945.

12

IRREGULAR WARFARE

Readers may think they know a lot about irregular warfare. They may be right. That is largely a twenty-first century phenomenon. Before the 'Global War on Terror', and military operations in Afghanistan and Iraq, irregular warfare was generally poorly understood. Veterans of those conflicts will, however, typically have seen a segment of just one, or two, of the 30 or so major irregular conflicts since 1978. And, of course, experience from Iraq and Afghanistan is not relevant to a study of the twentieth century.

Irregular warfare is grossly understudied. Of the 911 articles related to the twentieth century in 'War in History' between 1994 and 2016, just 10 referred to irregular warfare in any general way. The subject is studied elsewhere, but generally not by military historians. The Oxford Companion to Military History does not mention either civil war or insurgency.[1] COIN was described in just one of the thousand or so pages. That is as much as 'conquistadors', or 'military chaplains'.

Until the 1930s much of what is now called irregular warfare was often seen as tribal uprisings. Dealing with them was seen as an aspect of colonial policing. Overall, irregular warfare has been far more common than its regular counterpart. For example, the British Army lost soldiers killed in every single year of the twentieth century except 1969. Yet it was 'at war' for less than 20 years. Irregular war typically had far less consequence on a regional or global scale than regular war. It may, or may not, have had significant outcome in any one country; but typically only after several years, and rarely more widely. Unfortunately, domestic casualties were sometimes very high.

Irregular warfare occurs on land. Naval operations can make an important contribution, not least through blockade. Air force involvement

1 Richard Holmes ed, *The Oxford Companion to Military History*. (Oxford: Oxford University Press, 2001.)

has been far more visible. Its impact, as we shall see, has generally been highly overstated.

Chapter 12 considers definitions and then the occurrence and outcome of irregular conflicts. It explores the central paradox of irregular warfare. It then considers operational and tactical approaches, before making observations and deductions.

Definitions are human artefacts. Since conflict is fundamentally human and there is a wide spectrum of irregular conflict, we should not expect definitions in this area to be categoric nor exclusive. They tend to overlap. Exceptions abound. In this area some definitions seek to tie categories with political purpose or aspiration. That is unhelpful. Definitions can also be manipulated for the writer's benefit. Furthermore there is no 'definition police'. We cannot oblige different writers to abide by the same meanings. We can, at best, state a set of definitions and then be consistent in using them.

The simplest differentiation is between regular and irregular *forces*. That is fairly simple. Irregular *tactics* are mostly those of raiding and sabotage, by light forces which generally fight on foot. Those forces may be regular or irregular. Irregular, or guerrilla, *warfare* is an operational approach based on that. It may involve the creation (and hence defence) of enclaves. Irregular forces may expand their tactics to include using civilian vehicles, with or without weapons mounted on them, and locally-produced weapons.

Irregular forces often emerge during an uprising. An uprising is a collective resort to armed violence: an insurgency. Insurgents generally adopt guerrilla warfare techniques, described below. The use of government forces, foreign armed forces, or both to suppress an insurgency is COIN. It is generally an operational approach. That, and the relevant tactics, are considered below.

A militia is an irregular armed force. Militias can be pro- or anti-government. So some are insurgent militias. Some are no more than criminal gangs.

The general definition of war used in this book is 'collective armed violence for political purposes'. Collective armed violence *for the control of the state* generally involves both insurgent and pro-government forces. Regular armed forces may fight on one side or both. One term for such a conflict is 'civil war'. Civil war is not a subset of interstate war. It is a parallel phenomenon. Civil wars typically involve much of the country's armed forces fighting on the side of the *de facto* government.

Terrorism is a tactic. It is the use of violence to create fear for political purposes *by non-state actors*. It is generally a tactic of weak organisations. Some such organisations may be exclusively terrorist. Many insurgent groups employ terrorism as well as raiding, light infantry-type tactics. That is generally called 'guerilla warfare'.

The use of violence to create fear for political purposes *by state actors* is not 'terrorism'. If perpetrated against the state's own population, it is repression. If perpetrated in other countries, it is state-sponsored terrorism. Writers often equate repression with terrorism out of a sense of moral equivalence. That tends to overlook the difference of purpose. It understates the immorality of repression. Surely, to a liberal conscience, the first duty of a state is to *protect* its population?

Subversion is the covert, illicit or illegal manipulation of political processes. Methods may include intimidation, bribery, infiltration, secret subsidies, false-front propaganda and deniable paramilitary operations. Subversion may be conducted on its own, or as part of other operational approaches in irregular warfare.

Collective armed violence *for financial gain* is gangsterism. It is often conducted by 'warlords' or 'drug barons', with their gangs or militias. Insurgent or terrorist organisations can seek financial gain in order to fund their other activities. There is thus a blurred line between criminality and sub-state warfare. Gangsterism can only persist in states where law and order is weak. The most effective recourse is generally adequate law enforcement.

All conflicts are asymmetric to some extent but how much, and in what respects? It is doubtful whether the nature of 'asymmetry' *per se* is a useful subject of academic debate. Analysis which investigates disparities of ends, ways and means might be more productive.

We can identify two thresholds among the above definitions. The first is the choice to resort to armed violence. That is, to take up arms (for political purposes). That, of itself, tends to result in terrorism. The second, higher, threshold is the organisation of *collective* armed violence (fighting in groups). That is the point at which an insurgency can be identified. The distinction may not be clear-cut. In poorly-governed countries the population may live near to either or both of those thresholds much of the time.

The first threshold is nonetheless significant. It is, effectively, to kill for political purposes. The prohibition of killing is a strong taboo in practically every culture. It is the Sixth (or perhaps the Fifth) Commandment for the Abrahamic religions. There are at least two implications. The first is that the expression 'we had no choice' is inevitably a lie in this connection. One can always choose not to kill. 'We had no choice' actually means 'we would not actively consider other alternatives'.

The second implication is that the expression 'one man's terrorist is another man's freedom fighter' is vacuous. (The great majority have been men.) If the expression has any merit, it is to say that every 'freedom fighter' condones killing for political purposes. He, or she, may have valid reason for doing so. Yet the bland assumption of moral equivalence which 'one man's terrorist is another man's freedom fighter' implies is misplaced. It tends to ignore the fact that both terrorists and freedom fighters condone killing for political purposes. It also tends to *understate* the abhorrent nature of state-sponsored repression.

There is rarely a sharp divide between regular and irregular warfare. As we have seen, even the most 'regular' wars can have significant irregular aspects.

What can we observe about the occurrence and outcome of irregular warfare? An insurgency occurs where political issues have persuaded a significant faction or factions to resort to violence. If government forces succeed in suppressing an insurgency solely through military means, the insurgency will almost inevitably recur. Only two things will stop that happening: repression, or political change sufficient to redress the issue that caused the uprising. Repression will not resolve the underlying problem, which will fester (possibly for decades). The resolution of an insurgency will be political.

The word 'significant' (as in 'a significant faction') above is important. Extremists, or hostile foreign agents, may attempt to foster an uprising. Without popular support they will typically be neutralised by effective law and order measures. But in two cases that may not happen. The first is where law and order is insufficient to the task. The second is where their agitation identifies and exacerbates a critical political issue. Both cases will see the beginnings of an insurgency.

Insurgency and COIN should primarily be seen as a battle of ideas. It is a struggle for popular support. If the insurgents win, it will be because they have captured that support. Resolving an insurgency will always involve political change. (If the insurgents succeed, they will take power.

That is political change.) Cynically, any political change might be largely presentational. Political change will often involve new legislation and changes to systems of government. It will typically require economic development and social change, such as the provision of education and health care.

Much rubbish has been written about the duration of insurgencies and COIN campaigns. It tends to confuse correlation with mechanism. It may be entirely possible to change the military situation quite quickly (but see below). At best, that creates the conditions for political, hence social and economic, development. Those developments will probably take longer. However the protracted nature of many COIN campaigns resulted largely from a failure to make the necessary political changes in any timely fashion. Consensus may be hard to reach. Politicians may prevaricate or not agree to change. When such things happen, it becomes hard to sustain a military campaign which is largely limited to containing violence.

It is not clear how effective irregular warfare conducted by specialised military units has been. The evidence is ambiguous. Specially-trained regular forces conducting covert intelligence gathering, or raids, in regular warfare have had important successes. Examples include USMC Raider battalions in the Pacific, or the British Long Range Desert Group and SAS in the Western Desert. In the right circumstances the use of 'special' forces can have an impact out of proportion to the numbers involved.

Irregular (mostly British SAS) 'Scud hunting' operations in Western Iraq during the Gulf War in 1991 had an important strategic effect. They helped prevent Israeli involvement. The impact of the Chindits in Burma is far less clear. The operational impact of the two Chindit expeditions (in 1943 and 1944) was modest. There were second- or third-order benefits. Some were nebulous, such as 'moral ascendancy over the Japanese' (felt by who?) There were more than two divisions' worth of Chindits. There were never more than 10 other British and Empire divisions in Burma. It seems highly doubtful that the Chindits represented the best use of the available manpower.

About 30 per cent of uprisings since the Second World War have succeeded. An uprising is seen here to be typically a single (if protracted) insurgent campaign. The insurgents achieved some degree of success in a further 29 per cent of cases. Thus the insurgent had at least some success about 59 per cent of the time. However, the government achieved complete success in 41 per cent, and at least *some* success in 70 per cent, of cases.

Irregular warfare has often been strategically indecisive. Eight years after a civil war, a country has about a 40 per cent chance of returning to conflict. The fighting did not definitively resolve or settle the issue. The proportion decreases by only 1 per cent or so per year. In most conflicts resolution will be political; not military. Where political resolution is ineffective or insufficient, the country will revert to violence in a few years.

The organisational response to failure can be enlightening. The case of Vietnam suggests, firstly 'let's not do this'; then an aversion to casualties in subsequent conflicts; and a dislike of media attention. It can also lead to an organisational rejection in which the armed forces delete the 'lessons learnt' from their files. Regrettably, those who ignore the lessons of history are fated to repeat them.

The wars of decolonisation tell us something different. The colonial government may have been quite capable of achieving its intended strategic objectives. However, any military lessons may have become irrelevant after decolonisation came to an end. Alternatively they may have been hubristic, if the country (and its armed forces) think it can apply those lessons in different circumstances, decades later.

For the government, overall success occurs if the insurgent can no longer take recourse to violence. That is unlikely to be clear-cut. Insurgents will typically just drift away. COIN forces erode support, deny success, prevent recruitment, reduce the insurgents' freedom of action, interdict operations, capture arms, and kill or capture insurgents. That will not tend to be dramatic.

Consideration of COIN campaigns and tactics indicates a major apparent paradox. It is that some irregular forces have persisted, and continued to operate for many years, despite the activities of seemingly more powerful security forces. Some have eventually won. That seems paradoxical.

In practice, the paradox is easily resolved. Several insurgent groups have existed in what can be described as an evolutionary niche. It is a niche in the sense that the security forces cannot, in practice, bring their strength to bear. It is evolutionary in that the insurgents adapt and evolve to exploit local conditions for their own protection.

Those conditions may be partly geographical. Many insurgent groups have been able to exploit sanctuaries in places where government forces can not in practice engage them. Some sanctuaries lie across national boundaries. In other cases the terrain provides refuge; perhaps including forests, jungles and mountains.

The niche conditions may be partly social. In many countries, ethnic or other social divisions lead sections of the population to hide and protect insurgents from security forces. Some of those divisions might be denoted by religious affiliation.

Economic aspects may also be important. Some governments simply cannot afford protracted COIN operations (in which case there is no paradox). In some cases the insurgents have access to funding which allows them to continue long after they would otherwise have given up. Drug crops, or subsidies from other countries, may provide the funding.

There may be legal aspects to the niche. Tactically, many terrorist suspects in Northern Ireland remained at liberty because the security forces would not arrest them. Operationally or strategically, that was because legislation was not changed to support their conviction.

In Northern Ireland the problem was not a lack of intelligence. Sometimes terrorist suspects could not be convicted without revealing evidence, or sources, which the government or security forces were not prepared to reveal; or for lack of witnesses willing to testify; or similar. Changes to the law might have overcome that. There was legal conservatism, a commendable reluctance to infringe human rights, a lack of understanding of the dynamics of the problem, and resistance from lawyers with Republican sympathies. But, more importantly, there does not seem to have been a recognition of the nature, scope and scale of the issue. Had there been, there might have been a reasoned debate as to how to resolve it. That could have addressed ethical, legislative, security and jurisprudential concerns. Tragically that debate did not really take place.[2]

In practice every evolutionary insurgent niche will be different. Political, legal, social and economic conditions in every country are unique. Each insurgent force will adapt and evolve differently, not least because each faces a unique array of security forces. Those security forces will adapt and evolve in a unique way, and so on. The precise

2 I wrote the British Army's initial campaign study for Operation Banner, its military operations in Northern Ireland 1969-2006. *Operation Banner. An Analysis of Military Operations in Northern Ireland*. Army Code 71842, July 2006. To do so, amongst other things I read every historic document retained by HQ Northern Ireland and the British Ministry of Defence.

conditions of a given evolutionary niche will be a particular balance of geographical, political, social, legal and economic factors. Analysts and security forces should therefore be very cautious about applying templated solutions to insurgencies.

If a niche does not form the security forces will be able to prevail, and possibly quite quickly. That is, in perhaps a few months. That begs the question as to how government forces should respond. A list of empirical 'do's and 'don't's is given below. Another approach, however, would be to investigate quite explicitly what prevents the security forces from prevailing. That is, how to close the niche.

The immediate answer might be, for example, a lack of HUMINT but the underlying reason for that might be social (that the population won't betray the insurgents). The government and security forces might then address the reasons as to why that is the case. Alternatively, if geography is a major factor then better surveillance, improved mobility and better tactics may reduce or eliminate the niche.

If the answer is political (that the government cannot, or will not take, sufficient action) then analysis has reached the real core of the problem. Foreign intervention might affect the situation (for example, by invading or imposing a regime change). Yet without substantive political change the situation will not be resolved definitively. Remember also that Mrs Thatcher undertook no political initiatives in Northern Ireland. Between 59 and 125 people died in the Troubles each year that she was prime minister.

We now turn to the *methods* of irregular warfare. We start with the operational approaches used by the insurgents. Insurgent and terrorist organisations were often strategically naive. They had little idea how to apply collective armed violence purposefully. That has sometimes been unimportant. Insurgents typically seek to gain political power in a single, albeit protracted, campaign. They may conduct that campaign very well. Sometimes the campaign has been poorly planned, or executed, or both.

Insurgency can be a good way for a weak faction to gain power, particularly from an occupying power. Conversely terrorism has rarely worked by itself. Terrorism often seeks attention, acknowledgment and recognition. Those goals were often achieved. However, where they extended to authority and governance, they were almost never successful. That typically required a transition to insurgency.

This book does not consider political action which has no violent component. (That is, by definition, not war.) It is the admixture of (collective) violence which interests us. Insurgent groups have five main areas of activity. They are: persuasion; subversion; (violent) coercion; self-protection; and armed attacks on other groups, particularly government forces or institutions.

The causes of the uprising are fundamentally political. Many relate to direct power: the question of who shall run the country, or similar. Political power may be an aim in itself or may be related to the furtherance of social or economic issues. Religion is never a cause. At most, as in Iran in 1979 the issue was that a certain group of priests should replace the Shah's government. That is a simple issue of political power: the ability to govern. It had nothing to do with theology, nor of confessional differences. In Northern Ireland the Catholic hierarchy (priests and bishops) never publicly supported Republican terrorists. Furthermore the latter were almost exclusively secular in their political stance.

Typically only a small proportion of a population have been committed insurgents: perhaps 10 per cent. Another 10 per cent or so might have been committed to the government. The real struggle in an insurgency is for the support of the middle 80 per cent or so. Faced with real hardship, that 'trapped middle' will tend to side with strength, not moral virtue. Individuals tend to be influenced by the real practical issues of the moment: perhaps poverty, unemployment, duress or injustice. The main task for the insurgents is to persuade, subvert or coerce the uncommitted middle. The equivalent issue for COIN forces is to promote and maintain the loyalty of that uncommitted middle to the government. Simplistically that can be described as 'hearts and minds'.

COIN strategies are those of either indigenous or non-indigenous governments. The more significant case is that of non-indigenous parties, hence an 'intervention.' Interventionist governments, and their armed forces, have tended to overestimate the ability of armed forces to pacify but underestimate the importance of other, non-military activities. They also tended to underestimate the duration and the social, economic and financial costs. Interventionist governments can rarely separate the conduct of the intervention from domestic political issues in their homeland. Similarly for issues of prestige, credibility and international standing. Success is doubtful.

Good COIN Practices:	Bad COIN Practices:
Good strategic communications.	Collective punishment and escalating repression.
Significantly reduce tangible support to the insurgents.	The COIN force was an external occupier.
Establish or maintain legitimacy in the area of the conflict.	Security force or governments actions contributed to new grievances claimed by the insurgents.
The government was at least partially a democracy.	Militias worked at cross-purposes to the security force or governments.
Adequate intelligence.	Resettlement or removal of civilians for population control.
Security forces of sufficient strength to force the insurgents to fight as guerrillas.	Collateral damage by the security forces was perceived as worse than the insurgents'.
The government was competent.	The security forces were perceived locally as being worse than the insurgents.
Avoidance of excessive collateral damage, disproportionate use of force, or other illegitimate use of force.	The security forces failed to adapt their tactics, operations or strategy.
The security force sought to engage and establish positive relations with the local population.	The security forces engaged in more intimidation or coercion than the insurgents.
Short-term investment, infrastructure or development, or property reform, in the local area.	The insurgent force was individually more professional or better motivated than the security forces.
The majority of the local population supported the security forces.	The security forces relied on looting for sustainment.
The security force established and then expanded secure areas.	The security forces and government had different goals or levels of commitment.
The security force had, and then used, uncontested air superiority.	
The security force provided or ensured the provision of basic services.	
The security forces created or maintained a perception of security.	

Figure 12-1. Positive and Negative Practices for Success in COIN

Once an insurgency has become established it will not be resolved by military means. Hence the importance of (for example) the first 100 days after a regime change. Security forces should act quickly to prevent an insurgency breaking out. And even in that period, the key task is to gain the support of the majority of the population.

We know what wins. In 2010 analysts at Rand identified a series of positive and negative practices in achieving success in COIN.[3] There are 15 'good' and 12 'bad' practices, paraphrased in Figure 12-1.

The practices were identified in a study of 30 insurgencies from 1978 onwards. Success depended on the arithmetical sum of positive and negative factors. That is: do more good things than bad, and you will succeed. A positive sum (more good factors than bad) perfectly predicted success for the government side. A negative sum perfectly predicted failure. It is that simple. Simple, however, is not the same as easy.

Military operations will not be decisive. They will not resolve or settle the issues which drove the insurgents to violence. COIN forces *can* reduce the level of violence to allow political, social and economic development. They might persuade insurgents to desist because it is no longer effective, or cost-effective, to continue. Reducing the general level of violence in order to allow development is critical.

Whoever controls the ground can control the narrative. Control of the air is irrelevant to that. Control of the air did *not* give decisive advantage to COIN forces. The evidence before 1939 is, at best, ambiguous. The evidence from after 1945, of more than 100 campaigns, is not. As in regular warfare, air support was very useful for reconnaissance and intelligence gathering. It was hugely useful for tactical air transport.

Helicopters revolutionised COIN, but may have prompted a perception that COIN forces can come and go at will. Many airmobile operations in Vietnam only lasted a few hours. Arriving, staying and defeating the insurgents might have been better.

Air attack has had a very mixed impact in COIN. It can be helpful in support of ground troops in contact. That can be counterproductive when its use becomes the default tactic. Air attack can be very popular with the domestic population of an interventionist COIN force. It looks dramatic on

---

3 Christopher Paul, Colin Clarke and Beth Grill, *Victory has a Thousand Fathers*, (Santa Monica: The Rand Corporation, 2010), 88-89.

TV. The consequences are rarely visible to the home audience. It rarely has a cost in terms of lost aircrew. It can divide host-nation and non-indigenous authorities, because of collateral casualties and damage.

Air attack in COIN is only really effective where the insurgents can be made to stand and fight. That is rare. The damage may be transient. The resentment of collateral damage is generally not. Air attack is typically an ineffective substitute to the real business of defeating insurgents and bringing about political, economic and social change. It tends to be largely unnecessary, only effective in limited circumstances, and costly. Air attack has made very little difference to the outcome of COIN at the campaign level. We should, nonetheless, expect its supporters to continue to overstate its effects and underplay its failures. They have done so for decades.

We now turn to land force tactics and, again, start with the insurgents. Terrorists and insurgent groups are initially weak in military terms. They invariably start with raiding: usually ambushes and sabotage. As they progress, they typically attempt larger-scale operations. At some point they will generally attempt to hold ground, as enclaves. Before those enclaves are secure they often attempt to hide among the people. As they become stronger, the insurgents will attempt to attack and destroy government installations. They will typically seek more and better weapons, often from foreign sources.

Information operations – 'propaganda' – is an essential part of their armoury. It is their tool for persuasion, subversion, and one of the arms of coercion. It is a key vector for political action. Insurgent groups often have an advantage: their information operations are not necessarily constrained by the truth.

Insurgent groups often do not need to undertake economic or social development. They can simply blame the government for any perceived shortcomings. Local economic or social work can, however, help them persuade the uncommitted middle.

Funding is often a problem. Foreign governments have sometimes provided a lot of money. Insurgent groups have frequently resorted to robbery, especially where the targets (such as banks) can be associated with an unpopular faction. Insurgency is often associated with criminality, but the logic can be complex. Controlling the illicit drugs trade can be lucrative and prevent the emergence of rival gangs. However it can be socially unpopular.

Insurgent groups need to recruit in order to enlarge and to replace losses. They are therefore open to infiltration and betrayal. Security quickly becomes a major issue. The methods insurgents use to ensure security are often deliberately exemplary and gruesome. The need for security can dictate the shape and limit the size of the organisation and the tactics which it undertakes. PIRA's campaign in Northern Ireland is a prime example.

Most insurgent attacks are, at least initially, raids. It may be easy to achieve surprise. The insurgents can typically choose the location and timing of their attacks. It is difficult to prevent such raids directly. Opportunities for the security forces to achieve surprise and shock action will be rare and fleeting. Exploiting success will be very difficult. Thus armed action by the security forces is most unlikely to be decisive.

However, insurgents must raid. If they do not, they cannot commit violence and thereby achieve their objectives. Thus COIN tactics should largely be designed to deter or defeat raids and raiding.

Some of that will be protective security: observation posts, sentry positions, and so on. More offensively, patrols can create uncertainty and confusion among the raiders. However, they can also provide targets. So there can be an ongoing struggle at various levels. For example, numerous small security force patrols, using unpredictable routes, can swamp an area. That makes it difficult for the insurgents to operate freely. They will try to limit the security forces' freedom of operation, using ambushes and perhaps mines and boobytraps. Laying or setting them can provide more targets for security force patrols or ambushes. And so on. There will also be a technical struggle as (for example) more and better improvised explosive devices are developed, and counters deployed.

Security forces can benefit enormously from mobility. Helicopters provide a major one-sided advantage. They grant exceptional mobility, and particularly the ability to swamp an area with troops at little or no notice. However, as previously mentioned, they can create a belief that COIN forces can pacify areas without enduring presence. They can't. Once they fly away, the insurgents can infiltrate back.

There is no substitute for numbers. Enduring presence will require lots of security forces. In due course most (and eventually all of them) will be indigenous. They will have to be recruited, trained and supervised. They may become targets for infiltration by insurgents; an ill-disciplined rabble (which would be counterproductive); or simply another faction in the insurgency.

Identifying the insurgents and their plans will be a major intelligence challenge. Occasionally the product will be actionable intelligence which can be used to direct security force operations. That will tend to be infrequent, and rarely categoric. The results can occasionally be devastating.

Security force information operations ('government propaganda') were rudimentary in the twentieth century. Widespread broadcast media, and particularly digital media, only really developed in the 1990s. British attempts in Northern Ireland were initially seen as woeful and inadequate.

Overall, the most important requirement for a successful COIN campaign, or insurgency, was probably adaptation. No COIN force ever had the right organisations, tactics and equipment from the outset. Their opponents also often had to adapt, and hence evolve, simply to survive. So in one sense success went to the side which adapted best.

What can we observe or deduce? Irregular warfare in the twentieth century was widespread; understudied; but generally less consequential globally than interstate war. Individual conflicts were often protracted and indecisive: a country might fall back into conflict a few years later. Irregular warfare was generally indecisive both tactically (few engagements achieved anything) and operationally (fighting did not resolve the campaign). What the security forces *could* achieve by fighting was to suppress the insurgency. That would allow political, hence also economic and social, development to take place.

Insurgencies are resolved through political development. Where such development does not occur, the conflict *may* be halted by military action. It will probably break out again in a few years. That is because the underlying grievances, which caused a significant section of the population to resort to violence, were not addressed.

Insurgencies are not resolved by military means. Repression may well work, but will not resolve the underlying issue. The military mission should be to defeat the insurgents in order to allow political, hence economic and social, development. Defeated insurgents can no longer influence the uncommitted bulk of the population effectively. Here 'to defeat' means 'to prevent such influence'. Insurgency and COIN are the two sides in a struggle for the support of the population.

We know what wins: see Figure 12-1. We also know that military success may take a long time, because insurgent groups can occupy evolutionary niches where security forces cannot in practice bring collective armed force to bear. In each campaign the government and its security forces should try to identify what factors make up any such niche. Those factors will typically be a unique mixture of geography, law, sociology, economics and politics. Identifying the niche is the first step to forcing the insurgents out of it, or collapsing it around them. We should not expect that to work immediately. Insurgent groups adapt and evolve. Smart operational planning would ensure that the measures taken to eliminate an environmental niche are among the measures needed to address the underlying grievances. That is, the grievances which caused a section of the population to resort to violence. That would be both effective and efficient. It would also save lives.

13

Aerial Warfare

The early twentieth century saw the beginnings of manned, powered flight. That raised the possibility of war in the air. Aerial warfare then evolved in ways which Edwardian thinkers could only dream of. Perhaps the biggest of the many differences between warfare in 1900 and in 1999 relate to war in the air.

However, no other aspect of warfare has been so extensively politicised. Air forces have been the source of interservice wrangling over roles and resources ever since the Great War. Arguments about budget shares inevitably reflect assumptions and perceptions about roles. At its simplest, if air forces are thought to make the dominant contribution to winning wars, they should command the biggest slice of military budgets.

We will consider the term 'airpower' in due course. Until then we will simply refer to 'aerial warfare'. We will also refer to 'airmen' collectively. That is partly because very few women served in senior air force positions. It also reflects the absence of a less gendered term. This chapter looks at the nature of aerial warfare, and then at some of the institutional dynamics which accompanied it. That includes the rhetoric deployed around it. The chapter then arrives at a judgement as to how useful that has been, before making observations and deductions.

So, what is aerial warfare, and what is it not? What is it like? At its simplest, what writers describe as 'airpower' is the military use of the air. (We might have said 'the military use of the third dimension', but we must now exclude the military use of space.) Hence aerial warfare is the conduct of war in the air. Nothing more, and nothing less. The only importance of aerial warfare is its consequences for events on the ground; be they strategic, operational or tactical. That was identified in 1936[1] and remains

1 J C Slessor, *Air Power and Armies*. (Tuscaloosa, AL: The Alabama University Press 2009,) 61. Originally published by the Oxford University Press in 1936.

true today.[2] 'The conduct of war in the air' is the definition of aerial warfare used in this book.

The work of Colin Gray has been referred to already. It is important to understand that his work is mainstream airforce thinking, in the USAF and elsewhere. His 'Airpower for Strategic Effect' is widely used, for example, at the US Air University.[3] So in what follows we should take references from Gray to be entirely orthodox thinking about airpower.

Aircraft take off, do (or attempt to do) damage, and fly home. Aerial warfare is episodic. Logically it can only be classed as 'raiding'. That gives it peculiar and specific characteristics. Furthermore, in all warfare damage results from the employment of weapons. Damage is tactical. The conduct of air raids is tactical. *All* air attack is tactical in the first instance. As Gray put it, 'airpower itself … is and can only be tactical.'[4] So any discussion of aerial warfare reduces to whether the damage caused by aerial attack can, should, or does have operational or strategic effect. That logic is entirely clear. Almost everything beyond that, however, is assertion. The key questions are, firstly, whether the evidence supports such assertions. Secondly, whether the rhetoric deployed can in practice outweigh the empirical evidence.

We have seen that, in the Great War, German long-range bombers occupied a transient technical and tactical niche for about three months as they bombed London. Ironically, their greatest strategic impact was the creation of the RAF. Subsequently, long-range bombing did not win the Second World War. Major bombing operations did not resolve the Korean nor Vietnam Wars. They did not resolve the First Gulf War. They were important in resolving operations in Kosovo in 1999, but only as an instrument of political signalling. In short, long-range bombing has not had significant direct strategic effect. The only exception was in 1945 where bombing enabled atomic attack. Yet, to quote Gray again, bombing supported atomic attack; not *vice versa*.[5]

---

2 Colin S Gray, *Airpower for Strategic Effect.* (Maxwell Air Force Base, AL: Air University Press, 2012), 74.

3 Dr Heather Venable, personal communication. Venable is assistant professor of Military and Security Studies in the Department of Airpower at the United States Air Force's Air Command and Staff College.

4 Gray, 33.

5 Gray, 162.

Slessor was right in 1936.[6] He wrote that aerial attack has been at its most effective where it isolated the battlefield from reinforcement and supply. Slessor did not address naval warfare. In practice aerial attack at sea has also isolated *naval* engagements from reinforcement and supply. It has also struck to resolve naval engagements. Slessor could not say whether aerial attack was more effective in supporting land or naval operations. That is perhaps not a sensible question.

Slessor was no radical. Hugh Trenchard, the 'Father of the RAF' and CAS from 1919 to 1930, was not particularly articulate. He relied on a small group of staff to write key policy documents. He called them his 'English Merchants'. Slessor was one of them.[7] During the Second World War Slessor commanded No 5 Group in Bomber Command, and then Coastal Command. He later became the Chief of the Air Staff. Slessor wrote 'Airpower and Armies' in 1936 whilst he was the RAF instructor at the Army Staff College.

Essentially, Slessor considered that aerial attack is at its most effective where it supports surface operations, be they at land or sea. Aerial attack also made a highly important contribution to amphibious operations. It represents a significant conceptual difference between twentieth century warfare and anything which went before. We should no longer think of land operations, but rather air-land operations. Similarly for naval and amphibious operations. This conceptual distinction is highly significant. The concept of 'air exceptionalism' is the idea that the air is different from the surface of the earth; so aerial warfare is different; so air forces have to be different. However, as we shall see, aerial exceptionalism is misplaced.

Aerial warfare is conducted as raids. At the lowest tactical level, raids are attacks. Aerial warfare is (to that limited extent) intrinsically offensive. Slessor recognised two basic types of offensive: the attack on surface targets, and the secondary (but essential) gaining of aerial superiority. The former is the primary offensive because, to repeat, the only importance of aerial warfare is its consequences for events on the ground. So operations to gain and retain air superiority are essential, but secondary. In strict logical

---

6 Slessor, *Airpower and Armies*, 212.

7 Russ Mahoney, Trenchard's Doctrine: Organisational Culture, the 'Air Force Spirit' and the Foundation of the Royal Air Force in the Interwar Years'. *British Journal for Military History*, Volume 4, Issue 2, February 2018, 167.

terms, they are 'essential' because they are a necessary precondition to attacking surface targets.

'Bomber' Harris was perhaps the most famous (or notorious) British exponent of 'strategic' bombing. An analysis of how he planned and sequenced his attacks provides a very informative description of how to combine raids into major operations and campaigns.[8] He was well aware, for example, of the difficulties which arise from repeatedly raiding the same target. Air attack is essentially a raiding tactic, and raiding is rarely effective against strong defences[9] (like those which existed over Germany in 1943 and 1944). In early 1944, for example, Bomber Command's losses on raids against well-defended targets were, literally, unsustainable.[10]

Ironically, it was a highly decorated RAF navigator who was probably the first person to intellectually unpick the idea of 'strategic' bombing. Noble Frankland was awarded the Distinguished Flying Cross flying 34 bombing missions over Germany in the Second World War. He then gained a doctorate from Oxford. He concluded that 'the German records … left no doubt that the national effect [of bombing] on the German war effort as a whole was virtually nil.'[11] Frankland co-wrote the RAF's official history of the 'Strategic Air Offensive Against Germany'. There was considerable debate in the Air Ministry over its content and publication. There was also considerable public criticism of Frankland by former RAF commanders. Frankland's critique of long-range bombing is essentially a reflection on raiding as an operational approach. Identifying aerial attack as raiding is not new.

What else is aerial warfare, and what is it not? Like all warfare, it is adversarial and evolutionary. The observation above about the German long-range bomber force in the Great War reflects just that. In the Second World War the Luftwaffe and then Bomber Command found that they could not safely bomb in daylight. That was due to the strength of enemy ('adversary') defences. They adapted by moving to night bombing. That

8 Richard J Worral, ''Enough to be Seen to be Onside, but Hardly Substantial'? RAF Bomber Command and Operation *Husky*'. *War in History*, Vol 30 No 4, 443.

9 Jones, *The Art of War in the Western World*, 669.

10 Max Hastings, *Bomber Command*. (London: Pan Books, 1999), 267.

11 Obituary, *The Telegraph*, 24 November 2019. Noble Frankland, former Bomber Command navigator who revived the Imperial War Museum but ran into controversy with his official history of the air offensive against Germany – obituary (telegraph.co.uk). Accessed at 1515 hours GMT 31 January 2024.

prompted further counters, and so on. Thus aerial warfare evolved. Similarly, the USAAF found that it could not safely bomb in daylight until it developed long-range fighters. In a further development, the main effect of daylight bombing was the defeat of the Luftwaffe's day fighter force. That gained almost total air superiority over the land battle. And so on. (Aerial) warfare is adversarial and evolutionary.

Almost any kind of raiding *can* be devastating. Classically it works best when more-mobile raiders harry a broken, retreating enemy. The attacks by the RAF on the Turks at Megiddo, or the RAF and USAAF at Falaise, or Coalition air forces on the Basrah Highway in 1991, are all good examples. They are also good examples of ground-air synergy. To some extent the targets were 'set up' by ground force manoeuvre. That ground force manoeuvre had, in turn, been enabled or supported by air forces. Such synergy is both desirable and highly effective.

Raiding can also be demonstrative. Air attack excels at that. That is, air attack can be political signalling *par excellence*. The real virtue of the Kosovo air campaign was exactly that. Air attack showed Milošević that NATO was broadly united in its purpose; that military action was intended (and might result in a ground offensive); and that he should take political negotiations seriously. He did. He stopped attacking in Kosovo. (He then remained in power for more than a year.) Air attack as political signalling can be very useful to politicians. It can indicate intention at little or no risk of loss.

As we saw in Chapter 3, in 1917-8 43 German long-range bombers tied up 400 RFC and Royal Naval Air Service (RNAS) fighters. Raiding, and hence air attack, can cause a diversion of resources on the part of the defender. Unfortunately, the evidence of that working *well* in the case of air attack is ambivalent. The risk is that the attackers employ more resources than they divert. That seems to have been the case with the Combined Bomber Offensive (CBO). Even at the time scientists advised that, if diversion of enemy resources was the objective, it could be done better and at less cost.

Aerial warfare is intimately involved with technology. Its adversarial nature and the critical importance of small advantages in aerodynamic and weapons performance place a premium on cutting-edge technology. However the mere possession of high technology does not decide aerial warfare. Individual technological advantages tend to be transient. During any given war, their benefit has typically been short-lived: a matter of months at most. Again, the case of German long-range bombers in 1917-18 is a good example.

The effectiveness of raiding is related to the effects of individual raids. Clearly, in order to maximise that effect, the raids should be accurately 'targeted' (to use the modern meaning of the term). However it is now clear that warfare is not 'just targeting'. Modern targeting processes seek to maximise the effect of aerial attack; that is, air raids. But war and warfare *should* be more than just an accumulation of the effects of individual raids. If aerial attack is merely targeting then it will never amount to much. War is complex; precision, targeting and so on do not lead simply and directly to strategic outcome.[12] That is regardless of how much intelligence and technical precision goes into it.

The overall effect of raiding lies in the cumulative (and possibly synergistic) effect of the damage done in individual attacks. Yet any surprise and shock are transient, and damage can often be repaired or mitigated. Thus in aerial warfare the long-term effects have often been merely the questionable results of attrition. So, for example, the defeat of the Luftwaffe's day fighter force would have worn off, if the USAAF had not continued to fight to maintain it.

So much for the conduct of war in the air; but what about the institutions that wage it? The dominant or underlying issue has almost always been the independence of air forces.[13] That largely results from the concept of 'air exceptionalism'. 'Air exceptionalism' is perhaps the most significant issue related to war and warfare in the twentieth century. Is it helpful?

To an extent both the RAF and the USAF were misborn. The main motive for the formation of the RAF was for Lloyd George's government to be seen to be doing something about the German bombing of London.[14] A secondary motive was to create an institution which would give ministers leverage over the Army and Navy. When Trenchard (the designated Chief of the Air Staff) discovered that, he resigned.[15] The military value,

12 Gray, 236.

13 For example, Mahoney op cit. In 35 pages Mahoney uses 'independence' 24 times and 'independent' a further 13.

14 Gray, 98-9.

15 On 18 March 1918. That is, 13 days before the birth of the RAF.

or potential, of aerial warfare was less important than those other two, political, considerations.

The American case was different. In the Second World War the US bomber forces (the 8th, 15th and 20th Air Forces) enjoyed considerable autonomy. The Joint Chiefs of Staff considered creating a separate air force as early as 1944. It is not clear to what extent the Army wanted to divest itself of responsibility for air matters. Army Ground Force commanders probably wanted to get rid of involvement in long-range bombing, which they may well have seen as a distraction. That is, the USAAF had already won the argument, intellectually and emotionally. The US Navy (USN) wanted to keep its naval aviation. Army aviators wanted independence. The USSBS appeared to support the strategic effectiveness of long-range bombing (although its methods contain a fair 'degree of guesswork and statistical projection'.[16]) The role of atomic bombs in the surrender of Japan seemed to reinforce that. Army aviators won their part of the argument, and the US Army Air Forces (USAAF) became the (independent) US Air Force (USAF).

However, long-range bombing did not defeat Germany. The two atomic bombs actually showed that it was perfectly possible to drops atomic bombs from *Army* aircraft. Anything beyond that was inter-service political positioning. Those bombs *were*, nonetheless, hugely advantageous politically for the USAAF.

Thus, put simply, independent air forces were born under false pretences. Almost immediately after their creation they had to fight for roles, budgets, and even their independence. 'Institutional, reputational, budgetary and strategic challenges are inalienable from airpower.'[17] That fosters a degree of institutional insecurity.

Some airpower theorists, and senior airmen, have repeatedly over-claimed for the effectiveness of air forces: a process which started in the 1920s. As a direct result air forces have repeatedly under-delivered on such claims. They have then generally moved on successfully, without real consequence. Probably the best example of that is that many people still think that the CBO defeated Germany in the Second World War. 'Airpower

16 Hastings, *Bomber Command*, 324.
17 Gray, 101.

nearly always has been ... less decisive ... than its most impassioned practitioners have claimed.'[18]

Perhaps surprisingly, Gray wrote:

> '[I]t is asking too much to expect soldiers and sailors ... to understand airpower properly.'[19]

Thus airmen have effectively invented a discipline ('airpower') with its own philosophy and lexicon, which soldiers and sailors seemingly cannot understand, and which justifies the existence of independent air forces. That is exceptional, in several senses. Not least, there is no real naval nor land-force equivalent. In the 21st century it extended to the provision of masters'-level degree courses in airpower studies

Claims of strategic effectiveness for long-range aerial bombardment are not the only examples of over-claiming. Intellectually such claims arise from induction (the identification of the general out of the particular). Induction is philosophically unsafe.[20] It is, essentially, over-extrapolation. Two outstanding twentieth century examples come from the work of USAF Colonels John Boyd and John Warden.

Boyd extrapolated from credible theories about air combat manoeuvring by individual fighter aircraft (at the lowest tactical levels) to his famous 'OODA Loop'. From there he suggested theories of theatre and even strategic manoeuvre. They fail even casual inspection by reference to observed fact.[21]

Warden developed a simplistic 'bulls-eye' or 'onion skin' model for what he considered to be strategic attack. He proposed that long-range aerial bombardment should strike the inner circle of enemy power, the national leadership, as its strategic centre of gravity. Warden's main extrapolation was the assumption that long-range bombing inevitably has strategic effect. It doesn't. Bombing is tactical. Bombing power stations in the enemy's national capital, for example (which appeared as 'strategic' targets in both Iraq in 1991 and Yugoslavia in 1999) does not create significant strategic effect.

---

18 Gray, 19.
19 Gray, 309.
20 Karl Popper, *The Logic of Scientific Discovery*. (London: Hutchinson, 1959), 27.
21 Gray, 208.

To repeat: *does* the resulting damage have operational or strategic effect? Well, Iraq and Serbia were scarcely the world's most powerful nor rich countries. Western, Warden-inspired, USAF-led 'strategic' bombing in Iraq in 1991 and Serbia in 1999 did not produce strategic collapse. So even in cases of massive strategic overmatch, air attack did not have much strategic effect. Warden's theory is a classic example of over-extrapolation.

Both Boyd and Warden were airmen. Both produced theories which benefited air forces. Their theories were widely believed and taken into doctrine, both in air forces and more widely. Both theories are now generally discredited.[22]

In 1936 Slessor could not understand why two Great War airmen (Major General Frederick Sykes and Brigadier-General Percy Groves) had used the word 'strategic' for attacks on industrial infrastructure.[23] He thought the usage 'obscure'. However within a few years RAF and USAAF commanders were deliberately using 'strategic' as a euphemism for 'long-range'. USAAF doctrine actually described the relevant tactic as 'High Altitude Precision Daylight Bombing'. (The results achieved by the USAAF, in practice, were sometimes just as inaccurate as the RAF's. The USSBS went to considerable lengths to demonstrate the opposite.[24]) The term 'strategic' was employed for institutional politics, in order to emphasize long-range bombing's (and air forces') place in the debate over resources.

It endured. It still does. A Nimrod surveillance aircraft was described as 'strategic' by the then CAS when it crashed in Afghanistan in 2006. Did he actually mean 'long-range', 'scarce', or merely 'expensive'? Long-range bombing is long-range bombing. Bombing and the resulting damage are tactical. 'Long-range' does not mean 'strategic'. Using the term 'strategic' in this way is now largely unthinking, but actually reflects institutional politics.

The term 'strategy' can itself be ambiguous. In this book we define strategy narrowly as the conduct of war at the national level. However there is also a second meaning: 'the art of planning or directing military

22 Gray, 202-9. Also Stephen Robinson, *The Blind Strategist. John Boyd and the American Way of War*. (Chatswood, NSW: Exisle Publishing Pty Ltd, 2006), 51-71.
23 Slessor, *Airpower and Armies*, 69.
24 Hastings, *Bomber Command*, 324.

activity'.[25] That equates to 'just about anything written about warfare'. That gives plenty of licence for ambiguity.

Is anything related to aerial warfare 'decisive'? To repeat: something is decisive when it settles or resolves an issue. Logically there is no reason why aerial warfare cannot be decisive: tactically; operationally; or strategically. One can observe many instances of air fighting that have been decisive *within air combat*. That extends up to the defeat of the Luftwaffe day fighter force by the USAAF. That had major operational consequences. Did that, of itself, settle or resolve either the Normandy Campaign or (more widely) the Allied campaign in northwest Europe? No.

In practice aerial warfare has often had significant operational and sometimes strategic *impact*. Operation Allied Force clearly had some impact on the outcome of the Kosovo War of 1999. Did it, however, resolve or settle the issue? No.

'Having impact' is not the same as 'being decisive'. The word 'decisive' is bandied about without thought as to its actual meaning. It can then be misused. One claim for airpower was that, in one instance, its use was 'partially decisive'. Regrettably 'decisive' is binary. Something either settles or resolves an issue, or it does not. Something which is 'partially decisive' is actually 'contributary'. So, did aerial attack *contribute* to resolving an issue? Yes: that is entirely credible.

We then turn to 'effectiveness'. It is massively and repeatedly abused in connection with aerial warfare. 'Effectiveness' relates to outcomes. It does not relate to activity, nor output. For decades air force activity has been described in terms of numbers of aircraft used; sorties flown; and bombs (or other ordnance) dropped. Those terms relate to *activity*; not outcome.

Sometimes, after assessment, we see statistics relating to 'damage done'. That is, for example, so many tanks or aircraft destroyed. Those are measures of *output*. But do those numbers have any significance in relation to the wider battle? Or the campaign? If so, what *is* that significance? How do those outputs contribute to the *outcome*? Activity, output and outcome have repeatedly been confused. They still are.

Examples include the idea that all those bombs dropped on Germany *must have* had significant impact; or been decisive; or even had *strategic* impact. Regrettably that is not true. There is no 'must have' about it.

25 OED.

'Numbers of bombs dropped' is a measure of activity. Activity, output and outcome are not the same. They are not necessarily even closely related.

The pursuit and maintenance of independence led many airmen to believe that long-range bombing was essential, and then that it would or could win wars. That had grievous consequences. In both World Wars western armies and navies killed very few enemy civilians. Yet by the end of 1945 western airmen had killed about a million German and Japanese civilians. That is, well over 600,000 Germans and over 300,000 Japanese. Several million more were injured. Yet, sadly, today that is almost totally unremarkable.

At one stage in planning the bomber offensive, Portal estimated being able to kill 900,000 German civilians.[26] In the early stages of planning the notorious Dresden raid of February 1945, his staff considered the possibility of killing 110,000 civilians,[27] and thereby create what was seen as a positive strategic outcome.

The late 1940s saw revision of the Geneva Conventions, seeking to ban indiscriminate bombing. Yet, somehow, between 1939 and 1945 western air forces and their governments had been drawn into industrial levels of slaughter. Some of it was absolutely deliberate. 'Dehousing' was a euphemism. It described deliberate attacks on German civilians. It was often done by creating firestorms in German cities. Firebombing Japanese cities was done in the full knowledge that each of the dozens of raids would burn thousands of people to death. One historian has made a strong case that it was largely done to advance the cause of independence for the USAAF.[28]

Strategic planning in the Second World War was skewed for reasons related to the cause of air force independence. That extended down to operational levels. In Korea, Vietnam, Kuwait and Kosovo independent air forces sought to undertake largely independent air campaigns. They achieved little of significance. They were astonishingly costly. They probably diverted resources from other lines of operations, directly or indirectly.

Lack of direct effect is one problem. Lack of synergy is another. It is probably at least as important. The key factor is that independent air forces do not, institutionally, collaborate with surface forces for greatest

26 Hastings, *Bomber Command*. 180.

27 Ibid, 301.

28 William W Ralph, 'Improvised Destruction: Arnold, LeMay, and the Firebombing of Japan', *War in History* Vol 13 No 4, 519.

effect. They say that their independence is essential to ensure that 'the air weapon' is used to its greatest effect. In practice the independence of air forces guarantees that the air weapon is *not* used to greatest effect. It is used to support or underpin independence.

It does seem that the USAAF cooperated more closely with the US Army during the Second World War than the RAF did with the British Army. At the time, however, the USAAF was not independent. As soon as it became independent, the USAF started fighting its own campaigns (in Korea and then Vietnam, in the first instance).

Yet independence had brought tensions decades earlier. In March 1918 the British First Sea Lord (Beatty) wanted to bomb the U-Boat bases on the Belgian coast. A few months previously he could have just ordered the RNAS to do it. In practice, he had to *ask* the newly-independent RAF to do it. The request was declined. The RAF (perhaps reasonably) claimed that supporting the BEF was, at the time, a higher priority. The Navy's priorities weren't the RAF's priorities. A few months later the Army asked for a major effort to support its attack at Amiens. There was no competing request from the Navy. Yet, as Slessor put it in 1936, the RAF failed to concentrate force adequately.[29] That is 'adequately' both by the standards of an experienced RAF pilot (Slessor), and by the standards which the German air service had achieved in March that year. The RAF's priorities were clearly not the same as the Army's.

One could go on. To be honest, such evidence is not proof. However it illustrates that an independent air force's priorities are not the same as its army's and its navy's. If air forces are separate institutions with separate institutional priorities, their operational priorities will coincide only rarely, and by chance.

In summary: the existence of independent air forces is counterproductive both operationally and strategically. Those are the levels which count. The remedy is obvious. What is needed are genuinely joint naval-air and land-air operations. That will only happen when independent air forces are disbanded, and their resources returned to their nations' navies and armies. Naturally, the education and training of naval and army officers would have to be revisited to make more of them sufficiently 'air-minded'. But what is more expensive: a bit more officer

29 Slessor, *Air Power and Armies*, 184.

education, or independent air forces which result in flawed operations and strategy?

So: what can we observe or deduce about aerial warfare in the twentieth century?

Aerial warfare is, fundamentally, raiding. It has the same broad characteristics as any other form of raiding. Warfare is warfare, wherever it takes place.

Aerial attack on surface forces has, at times, been tactically devastating. It has had a major impact on naval warfare. Indeed it can be said that aerial attack, in its broadest sense, is now the means by which naval forces strike.

However, both 'air exceptionalism' and the concept of 'airpower' are counter-productive. A century of aerial warfare is long enough to make adequate observations and draw the right conclusions. We should not allow 'air exceptionalism' to 'stress the nobility and downplay the legacy' any longer.

Gray wrote:

> '... the principal strategic problem with airpower has not been, and is not, its tactical and technological immaturity. ... The fundamental error is one of fundamental foundational vision.'[30]

He was entirely right. However, he was wrong to suggest that the problem was the failure to properly explore the nature of aerial exceptionalism. Air warfare is *not* exceptional. The idea that air warfare is exceptional is counterproductive. Aerial warfare is raiding. It should be seen as such. Aerial attack is a very useful adjunct to surface warfare. It can be devastating. Yet nations, and alliances, would be more effective if they abolished independent air forces. They should then reconsider the organisation, planning and conduct of truly joint maritime-air and air-land operations.

To conclude, to repeat Slessor:

> 'No attitude could be more vain or irritating in its effects than to claim that the next great war — if and when it comes — will be decided in the air, and in the air alone.'[31]

30 Gray, 198.

31 Slessor, *Air Power and Armies*, 214.

14

Naval Warfare

Naval warfare is the conduct of war at sea. By 'at' we mean 'on', 'under' and 'over'. Naval warfare displays considerable continuity from the past, but that can be overshadowed by focussing excessively on technology. Chapter 14 looks at some of the enduring features of naval warfare, then some aspects of technology. It addresses some misperceptions about naval warfare before making observations and deductions.

Perhaps the most enduring aspect of man's use of the sea is that ships cost a lot. Warships are particularly expensive. Ships take a lot of money, and a long time, to build. They have typically been designed for a working life of 20 to 30 years. Building up fleets has typically taken decades.

Long hull life introduces several dynamics. One is that naval planners tend to look to the longer term. That has several long-term consequences. For example, rapid building programmes can lead to redundancy (when ships are no longer needed in such numbers) or block obsolescence (if they all need upgrading or replacing at or about the same time. That is invariably awkward to manage.)

That is all slightly overstated. It also takes years to build fleets of tanks or aircraft. Building up armies or air forces also takes years. The nub of the issue for navies is the cost of, and time to build, *individual* warships. Related to that, relatively few are built. A dozen major warships would be called a fleet. A dozen or so tanks is just a company.

Shipbuilding costs can be a major factor in national budgets. The cost of individual Dreadnoughts, then aircraft carriers, and then ballistic missile submarines, invariably attracted parliamentary scrutiny. Naval planners often had to fight hard to get the fleets they wanted, or needed. That was one of several factors that led to close scrutiny of duty cycles. How long could a ship, or submarine, be on station? How long did it need to be in maintenance? Or training up for duty with the fleet? How did all that affect overall availability or fleet size?

The cost of the pre-1914 Dreadnought race is an interesting case. The 28 British pre-war Dreadnought battleships cost about £4 billion in 2017 prices. By comparison, the six British Type 45 destroyers, built from 2010, cost £6 billion. However the British economy had grown about six-fold. So £4 billion represented perhaps £24 billion in the 2017 budget. That year Britain's defence budget was about £47 billion. The British Grand Fleet of 1914 had cost a lot, although not as much as might be imagined. The real issue is that, in the prewar world, no other nation could in practice afford to build a fleet of that size. By 1914 Germany had stopped trying to compete.

The issues were different in the case of the US Navy (USN) after the Second World War. By 1945 the US had an enormous fleet, but it would run out of hull life around the 1970s. The key challenge was to migrate that fleet over decades. It eventually centred on nuclear-powered submarines (both hunter-killer and ballistic missile) and fewer, but larger and nuclear-powered, aircraft carriers.

Cost, and in particular 'sunk' costs (to make a bad pun) suggest a second meaning to the idea of the 'fleet in being'. Operationally the term refers to the idea that a fleet, once built, can exert a controlling influence without leaving port. More generally, it can do so without ever offering battle. The logic is simple. A fleet in being is a threat. It is a powerful force which cannot be ignored. In the Second World War much of the Royal Navy was committed to guard against Italian and German naval units which rarely left harbour.

There are obvious counter-arguments to the idea. Firstly, one could say that forcing those units to stay in harbour broadly kept the seas free for use by the Allies. In practice the Allies largely kept the German surface fleet tied up in harbour in *both* World Wars. The Allies could therefore largely concentrate on the submarine threat. Secondly, the threat of a fleet in being is not credible unless it is sometimes exercised. Exercising the threat puts the fleet at risk.

There is, however, also a strategic aspect to the notion of a fleet in being. Instead of the idea that a fleet can exert a controlling influence without offering battle, one can suggest that a navy can be so large that no other nation nor alliance can in practice overcome it. The twentieth century provides four good examples. The German Navy did not in practice successfully challenge the Royal Navy in the Great War. The combined German and Italian navies did not in practice successfully challenge the Royal Navy in the Second World War. At much the same time the Japanese

Navy did not, in practice, successfully challenge the USN in the Pacific. Then, during the Cold War, the Soviet Navy did not in practice successfully challenge the USN globally.

This may or may not be a law of nature; but it does appear that despite the massive sums expended, it was not possible to overcome a considerably bigger fleet once built. That might be considered to be the strategic meaning of the term 'a fleet in being'.

The roles of navies also endure. Historically, navies had five main operational (theatre- or campaign-level) tasks. They were: the strategic movement of troops; the acquisition of advanced bases; the landing of armies on hostile shores; blockade; and contesting the mastery of local seas. Each has both offensive and defensive aspects.

That list deserves some scrutiny. To illustrate: convoying is a tactic. It is typically a defence against raiders. In the Great War Britain introduced convoying to protect merchant shipping from German submarines, who were trying to effect a trade blockade. In the Second World War the Royal Navy convoyed merchant ships in the Mediterranean not, primarily, because the Italian Navy was blockading Malta (it wasn't). It was because the Royal and Italian Navies were contesting the use of the central Mediterranean. British possession, and retention, of Malta was critical to that.

In the Pacific in the Second World War the USN conducted all five operational tasks. The blockade was largely conducted by submarines and aircraft. American submarines sank about half of Japanese merchant ship losses. A further third or so were sunk by aircraft, half of them carrier-based. Most of the remaining sixth were sunk by mines laid by USAAF aircraft.

In the later part of the century the expression 'the landing of armies on hostile shores' had to be modified. The use of naval aviation and long-range (typically cruise) missiles bring us to a redefinition as 'the projection of force ashore'.

What won wars? If we observe that neither World War could have been won without the US Army, then the great enabler was the strategic, transatlantic movement of troops. However the question merits further examination. It has been suggested that the Soviet Union could not have survived if Britain had been knocked out of the war (in 1940).[1] That may

1 Review of Evan Mawdsley, *The War for the Seas. A Maritime History of World War II.* (New Haven, CT: Yale University Press, 2020). Tim Benbow, *War in History* Vol 30 No 3, 343-344.

mean two things. One is that the supply of lend-lease material to the USSR was critical. Several Soviet commanders said that it was. However, little of it came from Britain, and little was carried in British ships. The second possibility is that, if Britain had been defeated, there would have been no way to move American forces to Europe. There would have been no staging base. Without that, or the threat of it, the whole of the Wehrmacht could have been concentrated to attack the Soviet Union. As it was, roughly a third was held back to defend western Europe. That is an intriguing possibility. It emphasizes the importance of strategic movement of forces by sea.

The many and major technological developments at sea in the twentieth century may blind us to the idea that navies have *always* tended to be technologically innovative. Was anything in the twentieth century as significant as the move from wood to steel, and sail to steam, in the nineteenth? The adoption and evolution of technology at sea is, on reflection, also enduring.

Several new technologies were adopted, or first used, in the twentieth century. They include torpedoes, mines, searchlights, radio, radar, submarines and (not least) aviation. To that we should add the carriage and launch of long-range ballistic missiles in submarines. That provided a nuclear deterrent capability that has remained safely undetected for decades. (Incidentally, that has no precedent.)

It is easy to pick fault with the way that practically any one technology was adopted and blame innate naval conservatism. However, the reality is more complex. Firstly, technology does not come in a box, to order. It is difficult to shepherd scientific development and to then use the results for combat advantage. Some technologies are over-hyped. Some are underappreciated. Perceived and actual need can be different. Some technologies can meet unanticipated needs. On balance, apparent conservatism may reflect general uncertainty. It may be difficult to know what the real requirement is. There may also be a confusing array of alternative solutions.[2]

---

2 Review of Vincent P O'Hara and Leonard R Heinz, *Innovating Victory: Naval Innovation in Three Wars*. (Annapolis, MD: Naval Institute Press: 2002). Kendrick Kuo, *War in History* Vol 30 No 3, 339-341.

Professional officers of any service often cling to the things which they know to work. There is good reason for that. If an alternative doesn't work, the result may mean death and defeat. Does that make officers inherently conservative? Or simply pragmatic?

Naval conservatism may be one of several misperceptions, or myths, about navies. Some are relatively unimportant but tell us something about what people want to believe. One example is the idea that no US Dreadnought battleship was ever sunk at sea. Well, that might be narrowly true. Let us take it apart. There were eight US battleships in Pearl Harbor on 7 December 1941. That evening none of them were fit for sea. None would rejoin the fleet for four months. Two never did.

One (USS *Pennsylvania*) was in dry dock. One (USS *Arizona*) sank and was never refloated. The remaining six were all sunk. They were all refloated. However one of them *was* subsequently lost at sea: the USS *Oklahoma* foundered and sank as it was being towed back to the USA. Therefore seven *were* sunk. *Two* were total losses, sunk, due to enemy action. One was lost *at sea*. To say that none were ever lost, to the enemy, at sea is (narrowly) true but people who concoct statements like that are telling you a lot about themselves.

A more important misperception revolves around the fate of battleships. A few years ago a British general said that battleships were made obsolete by aircraft. Let us look at the facts. A total of four British and American battleships were lost to aircraft in the Second World War. They were all lost in the space of just four days (7th to 10th December 1941), to Japanese aircraft, under conditions of strategic surprise.[3] None had air cover. The British lost three other battleships: one to gunfire (HMS *Hood*) and two to submarines (HMSs *Barham* and *Royal Oak*).[4] Just seven Allied battleships were lost in total. So, why should those navies believe that aircraft had made battleships obsolete?

The USN lost 12 aircraft carriers in the war. Seven were lost to aircraft, or to a combination of aircraft, submarines and surface ships. The British lost eight carriers: one to aircraft, five to submarines, one to gunfire and one to accident. So the Allies lost twenty aircraft carriers, of which eight

3 Two at Pearl Harbor; HMSs *Prince of Wales* and *Repulse* off Malaya.

4 *Hood* and *Repulse* were, strictly, battlecruisers.

to enemy aircraft. Overall, the enemy lost twice as many carriers to enemy aircraft as it did battleships. So did aircraft make *battleships* obsolete?

They did; but the reason lies not in the threat to battleships, but in the effectiveness of aircraft. Battleships were very effective at destroying enemy surface ships. They were difficult to sink and very useful for shore bombardment. They were also very expensive to build, crew and run. However, after 1945 the western navies had few obvious surface competitors. So what would battleships fight? Conversely, aircraft carriers could strike at perhaps ten times the range. That is: hundreds of miles, rather than tens. That includes ten times as far inland.

In practice the Soviet Navy built 14 *Sverdlov* class cruisers between 1948 and 1959. The Royal Navy had not yet developed carrier-based maritime strike aircraft capable of countering such a threat, so it retained five fairly modern battleships until such aircraft (in the shape of the Blackburn Buccaneer) entered service. The last of the British battleships was then scrapped.[5]

A further misconception concerns the importance of major naval battles. Put simply, there weren't many and they didn't account for many ship losses. There was one naval fleet action in the Russo-Japanese war. There were eight in the Great War and 37 in the Second World War. All up, the total might be 50 in the twentieth century. However, in the Great War only about a sixth of warship losses took place in fleet actions. In the Second World War it was just one in eight.

What fleet actions could do was change the operational conditions. Jutland and Midway were good examples. At Jutland Jellicoe came close to destroying the German High Seas Fleet. As a result, it never tried to engage the Grand Fleet again. At Midway the USN reduced the Japanese numerical superiority in carriers in the Pacific to effective parity. That limited the Japanese fleet's freedom of action. It limited Japan's expansion eastwards. To that extent, Midway was the turning point of the naval war in the Pacific. Yet only eight ships were sunk at Midway.

What observations or deductions can we make about naval warfare?

5 Tim Benbow, 'Largely a Matter of Sentiment'? The Demise of the Battleship in the post-1945 Royal Navy'. *Historical /Research,* 19 July 2024. 'Largely a matter of sentiment'? The demise of the battleship in the post-1945 Royal Navy* | Historical Research | Oxford Academic (oup.com) accessed at 17.24hrs BST 24 July 204.

Firstly, what it is not. It is perfectly possible that the political issue at stake in a given war would be maritime in nature. There was, however, no instance of that in the twentieth century. Furthermore, no war was decided at sea. Several major naval campaigns were conducted. Several contributed enormously to strategic success. All those wars were actually decided on land. Naval warfare was not strategically decisive. It might have been. However, in 1911 the British naval theorist Sir Julian Corbett described the issue very succinctly:

> 'Since men live upon the land and not upon the sea, great issues between nations at war have always been decided - except in the rarest of cases - either by what your army can do against your enemy's territory and national life or else by the fear of what the fleet makes it possible for your army to do.'[6]

The high cost of individual ships is a major factor in naval warfare, to an extent not seen in other domains.

The main operational roles of naval warfare are enduring. They are: the strategic movement of troops; the acquisition of advanced bases; the projection of force ashore; blockade; and contesting the mastery of local seas. Each has both offensive and defensive aspects.

The development and adoption of technology is a major factor in naval warfare, as in other domains. Perhaps the greatest single development in the twentieth century was the creation of naval aviation. In 1999 the Royal Navy's capstone doctrine declared that 'air supremacy is a necessary precondition of command of the sea'.[7] Strikingly, that is the only text on that page.

There are several misperceptions concerning naval warfare in the twentieth century. It is reasonable to assume that some will persist, and others will be added.

---

6 Julian Corbett, *Some Principles of Maritime Strategy*. (London: 1911), 16.

7 *The Fundamentals of British Maritime Doctrine*, BR 1806. (HMSO,1999), 32.

15

War and Warfare in the Twentieth Century

This book asks what we can learn from war and warfare in the twentieth century. But what anyone can learn depends greatly on what they already know. So in this book we have not tried to identify lessons. We have made several observations and deductions. It is for readers to consider *whether* they have learnt from them, and *what* they have learned.

Chapter 15 starts with a broad overview of the main observations from the previous chapters. It then discusses what we understand about *war*, and then what we understand about *warfare*. It considers the *study* of war and warfare, and then makes some concluding remarks.

We start with the main observations from the previous chapters. In doing so we risk some overlap with the more general comments which follow them. We focus largely on comments relevant to the future conduct of war.

The division of war and warfare into strategy, operations and tactics seems useful. The making of strategy, and the planning of operations at the campaign and theatre level, should be fairly simple (as discussed below). Now, in the early twenty-first century, it seems to be grossly overdone (at least in the west). That might be because practitioners do not focus sufficiently narrowly on their task. We will revisit that later.

Warfare should focus very sharply on outcome. That is, beneficial gain, and primarily at the national level. In the end, that is what matters. Operations and tactics should be the means to that end. The one really unavoidable caveat to that is the need to focus on human losses: an issue of efficiency or economy, in the first instance. Practitioners should focus on losses for two reasons. Pragmatically, they are a loss, and losses should generally be minimised (not least because in this instance they affect capability). The second reason is humanitarian considerations. That may not be an issue in some cultures.

War is collective armed violence for political purposes. Politics is ongoing. Hence there is always some unfinished business. War does not resolve or settle all political issues, and no issue is resolved or settled for all time. To even attempt to do so, one must totally defeat one's adversary and impose a new form of government.

War and warfare are underdefined, under-quantified and under-theorised. Attempts to make theory based on induction are typically flawed and unhelpful. We shall discuss how to avoid that later in this chapter.

Offensive operations can be broadly divided into two approaches. We have described the first as 'conventional tactics', in which an objective area is seized and then retained. The second is 'raiding', in which the attacker seizes objectives and then moves on. Some naval warfare, some land warfare and all aerial warfare consists of raiding.

In practice naval operations have generally, eventually, supported the resolution of conflict on land. Air operations should be directed to support surface (that is, naval and ground force) operations; albeit at times indirectly. To repeat Slessor: air warfare is of no consequence except in so far as it affects the situation on the surface.[1] Independent air operations are a gross mistake.

Irregular warfare has often been protracted and indecisive. At best it can create opportunities for political developments to resolve the issues underlying the conflict. The protracted nature of much irregular warfare may reflect two things. One is the formation and persistence of a survival niche for the insurgents. That niche may be political, legal, social, economic or cultural. The second is a failure to undertake political change.

Regular land warfare is fundamentally a combination of violence and movement. It shows four principal features. Firstly, it seems to be most effective, and may also be most economical and efficient, when the overall focus is on movement rather than violence. Secondly, without a mechanism for shock action, warfare is often protracted and indecisive. Thirdly, dismounted combat tends to be exactly that (protracted and indecisive), largely due to the dominance of the rifled bullet. Finally, armoured warfare provides a mechanism for shock, surprise and exploitation which makes it both a battle and a campaign winner. Remember that almost all conventional land conflicts since 1945 have lasted less than 90 days.

---

1 Slessor, Air Power and Armies, 7. Also Gray, Air Power for Strategic Effect, 74.

What can we now observe about our understanding of war (as opposed to warfare)? We defined war as collective armed violence for political purposes.

There are three great truisms about war. It is fundamentally an act of policy, as per that definition; it is fundamentally human; and it is not determined. (That is, in war cause does not lead reliably and repeatedly to effect.) In the twentieth century war saw the increasing use of novel technology. That was the acceleration of a trend easily visible in the nineteenth century. War also saw an acceleration in the mobilisation of state resources. That was visible from the eighteenth century, if not before. Indeed, in most major regards war displayed continuity of trend. That is: evolution, not revolution.

From one perspective, the impact of war was limited. GDP, and effective GDP per head, grew enormously in the twentieth century. Britain's GDP per head, for example, grew six times in real terms but in terms of world ranking, not much had changed. Most of the top ten countries in 1900 were still in the top ten or so in 1999. The big losers were Austria and Hungary, largely due to the breakup of empire. (It is not clear whether their GDP *per person* suffered much, if at all). What is perhaps remarkable is not that Britain remained in the top five or so countries. It is that its major overseas territories in 1900 (India, Canada and Australia) were in the top 15 countries in their own right by 1999. That overall *absence* of major change occurred despite the two most destructive wars in history.

There was *some* change. The European empires had broken up. Many new states were established, mostly as a result of decolonisation. Decolonisation might well have happened without the two World Wars, but probably more slowly. The other major change was the creation of effective international institutions, mostly under the banner of the UN. The UN has its flaws. The world was, however, unquestionably a better place in 1999 than in 1900. The UN is very much a part of that.

So much for the global impact of (major) wars. But what of war itself? War is basically an assault on the enemy as a collection of human institutions. It is an interaction between human organisations. War is adversarial, highly dynamic, complex and lethal. It is grounded in individual and collective human behaviour, and fought between human institutions that are themselves complex. Technology plays a significant role in the conduct of war. War is not determined, hence uncertain, and evolutionary. Critically, and to an extent that we should never overlook, war is fundamentally a human activity.[2] That leads us to several significant observations.

Firstly, war is adversarial. Each party starts with goals, which form the initial purpose of the warlike actions they undertake. That *should* initiate a golden thread of purpose to unite grand strategic objectives to the actions of individual soldiers, sailors and airmen. It would be very difficult to over-stress the significance of that. Where that thread is broken, the actions of individuals become purposeless or even counterproductive. War, and its attendant losses, becomes futile. How often has that claim been made? Can those who order and direct war now avoid a recurrence? That is a pious hope.

Winning a war means gaining political benefit. There is no such thing as 'winning the war but losing the peace'. That means 'losing the war'. The American General Willian Sherman said that war's legitimate object is a more perfect peace. That means a better peace. That means an improved political position. The US conducted the latter part of the Second World War on what has been described as the 'Roosevelt-Marshall principle', 'that strategy should be conducted without regard to post-war political considerations'.[3] Let us seriously hope that that never happens again.

Winning the war does not (just) mean 'winning the fighting'. Winning the fighting is valueless if it does not mean gaining political benefit. Given that there *will* be human losses, fighting (and even military success) *is* futile if it does not aim to obtain political change, and particularly political gain.

That requires brutal self-discipline in grand strategic purpose. At times that may be easy. For the Falklands Conflict, the strategic purpose was simply to return the Islands to British rule. More generally, it may not be easy to sustain strategic purpose in a democracy across several years, several administrations, and in wars of choice. For example, the United States' strategic goals for Vietnam clearly changed through the presidencies of Kennedy ('caution'), Johnson ('engage') and Nixon ('withdraw'). Was that why the end result looks like defeat? And whatever the answers to those questions, what about the military strategic advice they were given? Those three presidents were served by four Chairmen of the Joint Chiefs

2 See Jim Storr, *The Human Face of War*, (London: Continuum, 2009), 56.
3 Wilmott, *The Struggle for Europe*, 776.

of Staff. A glance into the twenty-first century suggests changing strategic goals for Britain, America and other nations for Iraq and Afghanistan.

Such brutal self-discipline may not come naturally to elected politicians, let alone a series of them. Nor does it necessarily come naturally to any other sort of political leader.

To revisit an issue from the Introduction: 'winning' means strategic, operational or tactical success. That means achieving goals in their higher-level context. Losing is the opposite. The definition is clear, but the reality may not be. *In war, 'winning' and 'losing' are vague and subjective.* Military goals are typically secret, both at the time and perhaps thereafter. Goals may change (war is evolutionary). Commanders may lie about what their goals are, or were. They may lie about achieving them (or failing to). Then, if winning means military and (particularly) political success, 'victory' is not just a pseudonym. Victory is a declaratory political artefact. It is man-made (that is, an artefact). It is not 'victory' unless it is declared as such. And such declarations generally have political purposes.

Raiding is not well recognised as such in Western military writing. Its main characteristics were described previously. The use of raiding as a tactic deserves much greater thought. It has major limitations.

Armed forces do not just fight. They manipulate the use, *and the threat of the use,* of violence. As a broad generalisation, regular armed forces are adept (and may be proficient) at the *use* of violence. By default and due to their political origins, irregular forces can be adept at managing the *threat* of the use of violence. Regular armed forces should give much greater thought to managing the threat of the use of force, against both regular and irregular adversaries. Violence is compelling. The mere threat of it compels the opponent to react. Exploiting that threat should be a key consideration.

This book deliberately *chooses* to identify and define strategic, operational and tactical levels of war. There is no fundamental law that directs that, or any similar, scheme of definitions. Tactics, operations and strategy are human constructs. Their definitions vary. That is a major problem. This book chooses to identify and define strategic, operational and tactical levels; quite narrowly. Warfare is generally underdefined, so doing that should aid clarity. It seems useful for practitioners.

Strategic direction is one thing. Tactics are clearly something different. There needs to be a mechanism for translating strategic direction into orders for tactical commanders. Geography dictates that such translation should vary between theatres. Thus there should be a separate plan for each theatre. Both those statements recognise that there are things called 'theatres of war'. It seems most effective to have one commander and one plan per theatre. It also seems most effective to translate strategic direction into orders for tactical commanders at the theatre level. Hence the identification of the operational level of war. Hence what is done at that level is identified here as 'operational art'. No more, no less.

It is, and should be, that simple: by definition. It can, and should, be done by very few people, quite simply, and quite quickly. However what one observes of the late twentieth and early twenty-first centuries is that it is grossly overdone. It should not be. It need not be.[4] We should spend 'less time splitting hairs over the nuances of Clausewitz'.[5]

The broad outline of operational (or 'campaign') design should be familiar to any planner. It is to understand the environment; understand the problem; understand the (strategic) guidance or direction; and make a plan. Those stages should overlap or be iterated to some extent. Yet, for example, the Israeli Shimon Naveh's PowerPoint turned that into about 116 explicit questions.[6] They seem to be drenched in metaphysics and casuistry.

Finally, war is not determined. (To repeat: cause does not lead reliably and repeatedly to effect.) There are several consequences. The most important is that war is unpredictable. It will remain unpredictable. Looking forward slightly, we can make just one prediction about war. That is that its conduct, course and outcome will remain unpredictable.

War is not determined. Hence it is not determined by technology. Superior technology can give an advantage. That is generally transient. Hilaire Belloc wrote that 'whatever happens, we have got – the Maxim Gun, and they have not'. Sixty years later the adversaries (African irregulars) still didn't have Maxim guns. They had Kalashnikovs, when their opponents (colonial security forces) often only had bolt-action rifles. In general, numerical analysis has rarely (if ever) demonstrated any persisting military

4 M L R Smith, On Efficacy: A Beginner's Guide to Strategic Theory. *Military Strategy Magazine*, Vol 8 Issue 2.

5 William Thomas Allison, review of Donald Stoker, *Why America Loses Wars: Limited War and US Strategy from the Korean War to the Present*. In *War in History*, Vol 29 No 2, 518.

6 Shimon Naveh PowerPoint | PPT (slideshare.net) accessed at 11.51hrs GMT on 15 November 2023.

advantage due to a particular technology in war. Technological advantage tends to be transient.

War is not determined. Hence it is not determined by numbers. Numerical superiority can give an advantage. 'More Russians', for example, were more effective against the Wehrmacht than 'fewer Russians'. However the advantage is relatively small. Tactically the Russians were consistently much less effective, unit for unit, than the Germans.

War is not determined. Hence it is not determined by information. Information superiority can give an advantage. Numerical analysis shows that information superiority, of itself, is of little benefit. Its benefit lies in the specifics of what it allows the possessor to do. The most important is to achieve and then exploit surprise. Similarly, deciding and acting faster does not, of itself, require information superiority. It does have a significant impact on outcome.

Overall, war is not determined. What gives one adversary the greatest advantage are human factors. They include professional knowledge and skill; training; and leadership and motivation. War is fundamentally human.

Perhaps not surprisingly, the foregoing discussion on the nature of war has strayed into observations about how to conduct it. What more can we say about our understanding of war*fare*, the conduct of war?

Firstly, warfare is poorly contextualised. Writers and thinkers seem to have a poor understanding of war and hence warfare. One often reads that war is 'paradoxical' and 'ironic'. Logically a paradox is a pair of statements which both appear true, but cannot both be true at the same time. To resolve a paradox, one must investigate the underlying paradigm. A better understanding of that paradigm will show that the apparent contradiction is unfounded. That observation can be found in popular texts on philosophy.[7] It is not rocket science. Similarly 'ironic' means the 'expression of meaning through the use of language signifying the opposite'.[8] In this usage paradox and irony both indicate the same cause. That is that the writer did not actually understand the phenomenon being described.

7 Martin Hollis, *Introduction to Philosophy*. (Oxford, Basil Blackwell, 1985), 51.
8 OED.

So: where a writer indicates paradox, irony or both in connection with warfare, she or he does not actually understand what they are writing about. It would be hard to stress that too much.

As described above, *war* is adversarial, highly dynamic, complex, lethal, human, not determined, hence uncertain and evolutionary. That is the context. The *conduct* of war should reflect that. Briefly (and not exclusively):

1. War is complex. Mental tools such as the Three 'E's model (of Economy, Efficiency and Effectiveness) help understand how to manage complexity. Fundamentally, commanders should aim to resolve or manage complexity: not analyse it. Academics have wallowed in 'the complexity of modern war' for far too long. That has led many officers in the wrong direction.[9]
2. War is human. Warfare should focus on human behaviour, both individual and collective. That includes institutional dynamics. The issue of 'airpower' described in chapter 13 is the most egregious example of undesirable institutional dynamics to emerge from this study of the twentieth century.
3. War is not determined. Therefore warfare should proceed on the basis that outcomes and even outputs are not predictable. There is no point, for example, in writing plans and orders which rely on predicting the course or outcome of a given operation.
4. War is evolutionary. Armed forces must inevitably 'learn by fighting' to some extent. Warfare is a learning project.[10] The same applies to both sides. Critics of the conduct of (say) the Great War should consider what the major adversaries knew in 1914; how much they learnt; and how much their opponents learnt. Only then should they consider whether those adversaries should have known better.

Secondly, warfare is under-defined. In most intellectual disciplines, definitions are agreed and the list of definitions is sufficient to the task. Imagine trying to be a doctor without agreed definitions of 'medicine', 'surgery' or even 'patient'.

9 Storr, *The Human Face of War*, 52-3.
10 Grey, *Airpower for Strategic Effect*, 140.

In warfare there *are* common meanings for most phenomena. The problem is that 'common meanings' are typically loose and broad. There is ambiguity and overlap. We have seen that 'strategy' means at least two things. Those meanings are rarely discriminated between. That leads to overlap. For example, there is an apparent overlap (and hence disagreement) between 'operational art' and 'strategy'. Some writers seem to exploit such ambiguity wilfully. That is bad enough. Others seem unaware of the ambiguity, which is simply amateur.

As a test, ask a colleague to define 'command and control'.

We should not put up with such lack of precision. Some writers revel in it. One reads, for example, that "victory' is a contested concept'. Well, it shouldn't be. We should, and can, do better than that. We are now well into the twenty-first century. We have had staff colleges for centuries and university departments of war studies for decades. Regrettably there are no 'definition police'. This lack of precision will persist unless someone takes a stand against it.

Thirdly, warfare is under-theorised. There are few (if any) accepted theories of warfare in the sense of, say, the Newtonian theory of gravity. That significantly limits our ability to progress. One occasionally reads of the need for theories in warfare.[11] An example would be the need for 'a theory of victory' (sic). However writers usually seem unaware of what such a theory would look like or consist of. With due respect to the military historians, who probably form the majority of writers on warfare, we may be expecting too much. Theorising is not what they do.

There is, however, a more fundamental problem. The twentieth century gives several examples of flawed military theories. They include the work of Liddell Hart, Boyd, Warden, Mitchell, Douhet and others. The underlying problem is inductionism.

Inductionism is the intellectual process of going from the particular to the general. Liddell Hart provides an example. In 'The Science of Infantry Tactics Simplified', he effectively *described* best practice from the Great War.[12] Coupled to Fuller's 'Plan 1919' (which was essentially a concept paper), Liddell Hart then developed his 'expanding torrent' theory,

11 Jeremy Black, 'Where does Theory go in Military History?' *War in History,* Vol 29 No 1, 68.
12 B H Liddell Hart, A Science of Infantry Tactics. *The Military Engineer*, Vol. 13 No. 71, 409-414.

which effectively extended infiltration up to theatre level. That was gross over-extrapolation.

Inductionism is intellectually unsafe. It is reasonable, when the conclusions are close to the evidence to hand. That would result in 'cogent' theory. 'Cogent' means 'clear, logical and convincing'. It does not mean 'proven'.

Contrast the above with the German development of armoured warfare in the interwar years. In his staff studies, von Seeckt directed the formal collation of masses of empirical evidence. That was then used to revise doctrine. To that, add years of secret trials and development work. Panzer divisions were formed in 1936 and exercised repeatedly. The Wehrmacht then used those divisions in Austria, Czechoslovakia and Poland. They integrated the Luftwaffe into their tactics. They developed those tactics repeatedly through France, North Africa and Russia. So at every stage the tactics in use were based very closely on empirical evidence, duly assessed and analysed. That is an example of limited inductionism. The resulting practice was not over-extrapolation. There was little explicit theory. Westerners call it 'blitzkrieg'.

One might say something similar about Soviet deep battle theory. It did stand on the Soviets' experience in the Russian Civil War. The Soviets did conduct trials and large-scale exercises. So we can say that Red Army's theory of deep operations stayed moderately close to the empirical evidence.

That cannot be said of Liddell Hart. Nor Boyd, Warden, Mitchell, Douhet and others. It can be said of *one* airpower theorist: Slessor. Interestingly, Time Magazine stated that 'In 1936 Slessor wrote the pioneer text on strategic air bombardment', and 'then in the Second World War he used it.'[13] The first statement could not be much further from the truth. The second is therefore not true either.

Inductive theory may be unsafe but may be useful if we use it to move forward gradually. In other words, if we are sceptical. In warfare induction can be dangerous. Unsafe theory can cost lives, lose battles and perhaps even campaigns. Warfare *is* under-theorised. Any theory, however, should remain close to the evidence, such as it is. As we move into the twenty-first century and look back at concepts such as 'Effects-Based Operations', 'Network-Centric Warfare', the wider claims for cyber 'warfare', Artificial

---

13 Slessor, *Airpower and Armies*, 2009 edition, back cover.

Intelligence and now 'Multi-Domain Operations', we should pour in a healthy dose of scepticism. *What is the evidence*?

Theory can, nevertheless, be useful even if wrong.[14] Many established sciences have examples of theories that have been discounted. They generally advanced the discipline to some extent, even if perhaps in a misleading direction, before being discarded. In many cases, their apparent shortcomings prompted a search for a better theory. Thus progress *was* made. However, once again: military practitioners should proceed cautiously. The cost of error is high.

Fourthly, warfare is under-quantified. The history of warfare in the twentieth century has produced a vast amount of data. Some of it has been quantified, and the results analysed. That is how we know, for example, that the most important role of air support to land operations is reconnaissance of enemy ground forces. Quantitative analysis also tells us that in urban operations the defenders always suffer *more* casualties than the attackers. That is how we know that 'urban exceptionalism' is a myth and that many historians have simply been wrong.

The general absence of quantification has allowed wild claims, myths and unsubstantiated conclusions to persist. Some have persisted for 70 years in the case of the Second World War, and over a century for the Great War. That is now, slowly, being rectified.

Quantification can be time-consuming and expensive. Some of the statistical analysis is complex and difficult. Some things, however, can be explored through very simple numerical techniques. They include simply counting, or looking at published numbers critically. For example, how many panzer divisions did Germany have in 1944? How many (broadly equivalent) tank and mechanised corps did the Red Army have? How many armoured divisions did the western Allies get ashore that year? And what can we deduce from that?

We shall shortly see that historians are not generally inclined to perform numerical analysis. So we are typically missing an important part of the story. Of course there are 'lies, damned lies and statistics'. Therefore we must consider not just the numbers but their provenance, their relevance, and their significance. We should consider the numbers carefully.

14 Thomas Kuhn, *The Structure of Scientific Revolutions*. (London: University of Chicago Press, 1996), 104ff.

Warfare is under-quantified. More quantification is needed. That implies two things. One is that military historians would benefit from some training in numerical methods. Not least, that would make them more aware of the need for quantification. The second is that thinkers of a more mathematical (or simply numerical) inclination should be encouraged to engage with the study of warfare.

Lastly, the role of doctrine in warfare is poorly understood. An American officer once said that 'doctrine is the last resort of the mentally feeble'. It is a pity that he thought that. Yet in a sense he was right. If an officer can only argue an issue or viewpoint on the basis that 'it says so in our doctrine', then that officer's training and education are demonstrably deficit.

Doctrine exists at several intellectual levels. Low-level doctrine relates to fairly prosaic issues such as 'how to fire a rifle', but also 'how to write an operation order'. Higher-level doctrine relates to philosophy and principles. It is informative and educative rather than prescriptive and instructional. An example of 'prescriptive' would be 'this is the way that you *are to* fire a rifle', with the verb in the imperative mood.

There is, however, a more important (if subtle) distinction. Most doctrine is primarily normative. It seeks to establish or sustain standards (or norms) of behaviour. Sometimes that needs to be updated for broadly technical reasons. For example, when an army introduces a modification to its rifles, it needs to tell soldiers how to fire the modified weapon, and do so safely. Sometimes doctrine needs to be modified as a direct result of lessons learnt. For example, British infantry doctrine was modified after D-Day in an attempt to reduce officer casualties (as described in Chapter 8).

That should be contrasted with 'aspirational' doctrine. Change is thought to be needed, so doctrine is revised to describe and promulgate the intended changes. The previous example was partly aspirational (it was *intended* to reduce officer casualties). However aspirational doctrine can also be introduced to change practice without such direct, empirical evidence. US AirLand Battle is a good example. An attempt was made to introduce infiltration at and around the battalion level into the British Army in 1975.[15] It failed. It was repeated in 2005.[16] That also failed. The

15 Infantry Training, Volume IX. Pamphlet No 44 Part 1, *The Infantry Battalion (General)*, 1975. Army Code 70741. Chapter 9.

16 Army Doctrine Publication *'Land Operations'*, 2005. Army Code 71819, 70.

main reason for those failures seems to have been a lack of 'socialisation'. The change was not promulgated sufficiently well, and the target audience did not see the need for the change. Conversely, AirLand Battle was very well socialised. War, hence warfare, is fundamentally human. Such human issues often prevail.

In one sense even aspirational doctrine is normative. It seeks to establish *new* standards (or norms) of behaviour. When analysing published doctrine, one should identify what is new and consider to what extent that is aspirational. We should then consider the reasons underlying the aspiration. The results can be highly informative.

We now turn to the *study* of war and warfare. All intellectual disciplines have blind spots, heuristics and biases. For example, a study of academic economists in America found that they were three times as likely to give politically left-leaning opinions, rather than right-leaning. Examples would include attitudes towards wealth distribution, as opposed to wealth creation. For anthropologists, the ratio was 30 to one.

In practice most of the study of war and warfare is done by historians. They may be academic historians, or trained by them. We can describe some of those who are not as 'citizen historians'. Historians generally follow similar processes, most of which were formulated by academic historians. An informal survey of military historians suggested that they tend to display:

1. a strong preference for written sources;
2. a tendency towards discourse about written sources;
3. preference for the results of the above over information gained elsewhere;
4. blind spots resulting from gaps or absences in the above;
5. a disinclination to study maps in detail;
6. a disinclination towards numerical analysis;
7. an ambivalence towards technology, which might be either:

    a. a tendency of some to dwell, or even revel, in the small details of military hardware; or
    b. a tendency of some to resort to technological determinism; or
    c. an aversion towards technology by some others.

That may not be surprising. (1) and (2) can be seen as what historians do. (3) and (4) are natural consequences of that. We have noted several such gaps. (5) results from few of them being trained to do so (as most army officers, for example, are). (6) probably reflects the fact that very few academic historians have even A-level (high school) mathematics (a consequence of school structures). (7)(a) and (b) may be interesting to some readers. They are unhelpful to our purpose here. They result in part from the fact that exceptionally few people are actually trained as military technologists. (Many people know a lot about technical details. That does not of itself make them good technologists.) (7)(c) is also unhelpful. It may be a reaction from mainstream historians, who feel that (7)(a) and (b) give military history a bad name.

One highly respected British military historian remarked of his education at Cambridge University that it sometimes appeared that *how* students wrote was more important than *what* they wrote. Another history professor remarked whimsically that God is not so almighty that he can change history: therefore he created historians. Tongue in cheek, but perhaps more tellingly:

> 'Humans are the most unobservant creatures in the universe. Oh, there are lots of anomalies ... but historians explain them away. They are so very useful in that respect.' - 'Death', in '*A Thief in Time*'; Terry Pratchett.[17]

History is our best guide to the events of the future, but it is an imperfect mirror. The historical method revolves around the analysis of written records. In that, much excellent work has been done and will no doubt continue to be done. However, seen from a distance and looking at a century of military history, there seem to be two main problems. One is that historians do not seem to notice when something is missing. The other is that they tend to follow established interpretations.

The first problem is not easy to fix. Anaximander, born in the seventh century BC, wrote that progress is not made by accumulating facts, but by knowing what it is that one doesn't know.[18] This problem reminds us of Sherlock Holmes' 'dog that didn't bark'. It seems to take the skill of the

17 Pratchett, Terry, *A Thief in Time.* (London: Harper Collins, 2001), 66.
18 'Great Minds do not Think Alike'. *The Economist*, 11 February 2023, 76.

world's greatest detective to notice that something is missing, let alone deduce the reason or assess the consequences. In Chapter 7 we remarked five times that something was missing. Perhaps Pratchett was right.

The second issue is that of the 'dominant narrative'. Countless books have discussed the failure, futility and slaughter of the Battle of the Somme (for example). They include good books by respected historians who hold a favourable view of Haig. They tend to dwell on the events of 1 July 1916 north of the Albert-Bapaume Road. They may admit that some ground was captured south of the road but the maps, the ground taken, and the topography speak for themselves. The account given in this book may have shortcomings. However, the fact that significant sections of the German first, second and third positions were each seized in one day (on 1 July, 14 July and 20 September 1916) surely tells us something important.

This is not the place for in-depth historiographical debate. That has been avoided in this book, for reasons we will consider later. It seems, however, that historians often follow established interpretations. Such interpretations seem to be established quite easily. Once established, they are hard to undo. That may partly account for the problem of gaps and omissions noted earlier.

The structure of the humanities, or liberal arts, is fuzzy. History seems to be one of the humanities. Is it a social science? Historians tend to say that they are not scientists. Apart from history, most of the study into war and warfare takes place within the social sciences. Social scientists often approach war and hence warfare from international relations. They may employ the tools of social science research. The behavioural human sciences seem to lie among, or perhaps alongside, the social sciences. They are: psychology (the study of human behaviour), and particularly applied psychology (the study of human interaction); sociology (the study of human institutions); and anthropology (the study of human culture: typically myths, values, symbols and beliefs). Anthropology extends into archaeology (which can be seen as the study of physical artefacts; that is, physical culture).

We should note 'the general absence of theory and the linked poverty of the fall-back theoretical basket of the subject with such staples as War and Society, Face of Battle and Military Revolutions'.[19] Similarly, history

19 Jeremy Black, Where does Theory go in Military History? *War in History*, Vol 29 No 1, 68.

is light on theory, whereas social science is light on the history that is the only source of empirical evidence for theory.[20] So, social science is short on evidence and long on rationalistic discussion of the discourse.

War is ineluctably human. The behavioural sciences seem to have much to offer to the study of war and warfare. Unfortunately their contributions have generally been limited, although at times deeply insightful. Practitioners in some of those disciplines seem not to talk to those in others. In some cases studying war has been seen as a moral betrayal of the discipline. Methods differ. That can cause animosity. All that is unfortunate. There appears to be great untapped potential.

Such problems may be typical. Globally, the number of university researchers grew fourfold from 1980 to 2024. Yet in the west productivity growth fell from four per cent per year in the 1950s and 1960s to one per cent today. Productivity growth is largely due to innovation and development. It is reasonable to believe that this fall is due to a shift away from in-house research and development facilities in firms to universities.[21] So, we now have far more academics. They do a lot of work. That is: they produce a lot of out*put*. But the out*come* is greatly reduced. Put simply, universities do much less *translational* research.

The ultimate value of studying war and warfare is to practitioners: sailors, airmen and soldiers. That often seems to be forgotten. Two recent examples illustrate the problem. One was an international conference in 2022. 148 papers were to be presented. Speakers came from about 30 countries. Yet analysis of the programme indicated that not a single paper would benefit military practice in any useful way. It was, to be quite judgmental, an academic lovefest. Academics were writing for, and speaking to, academics. A few officers sat in the audience, probably thinking that somehow it was helpful to them. That seems unlikely.

The second example was a chapter published in a book in 2023. It was typical of the others in the book. It runs to 30 pages. It has 85 footnotes, which are all references. About 80 per cent are references to other secondary sources. The remainder are policy documents, or similar. In other words, it was 30 pages of discourse about discourse. It told practitioners absolutely nothing of concrete worth. All this is ultimately so pointless. The authors

20 Gray, 268.
21 Free Exchange: 'Ivory Sours'. *The Economist*, 10 February 2024, 69.

would, of course, say otherwise. In summary, much of the academic practice of 'war studies' has little or no value to practitioners. Yet it forms half of the syllabus in many staff colleges.

A colleague was appointed professor of the history of war in 2022. During questions after his inaugural lecture, he confirmed that generals (and presumably admirals and air marshals) did seem to listen to what he said. He was rather less confident that they did anything as a consequence. We can now see why. He had made a cogent case that, by the end of the twentieth century, academics knew far more than military men about the *history of war*. That had not been the case a century earlier. That, however, is not the point. Senior military commanders are not stupid. The history of war is *not* their business. Their business is warfare. They should be interested in the history of war*fare* and how that can inform their practice. From that perspective, many military historians are not addressing the right issue.

What overarching conclusions can we draw in relation to war and warfare in the twentieth century? The comments here should be considered in conjunction with those at the beginning of this chapter.

The biggest single observation is not surprising. Wars are won by better strategy, better operational art and better tactics. Better strategy includes generating the forces required in sufficient quantity and quality to do the job. But overall, of course, wars are actually won (or lost) at the strategic level. The only thing that really matters in the long run is the strategic outcome.

Furthermore, a war is not won unless a desirable political change is achieved. Anything short of that is not winning. To think so is naïve. In particular, winning the fighting but failing to achieve a desirable political change is not winning. It leads to futility and pointless deaths.

Most aspects of war (including its causes, conduct and outcome) are dominated by human behaviour, both individually and (more importantly) collectively.

Wars in the twentieth century were won on land. Aerial warfare has been best employed to support surface warfare. Naval warfare has been a critical enabler to strategic success on land.

Greater clarity of thought is needed. That should start with more attention to definitions. Furthermore, there has been plenty of history of war, but not enough history of warfare. In particular there has not been enough quantitative analysis of war.

There is a great need for more and better theory to support warfare. Such theory should remain close to the empirical evidence. Good evidence is hard to ascertain. For one thing, historical evidence is rarely categorically true, whatever 'true' may mean. Good quantitative evidence can be especially valuable.

Empirical evidence is hugely important. Urban warfare gives us one egregious example. Yet academics persist with ideas of urban exceptionalism despite categoric evidence that they are wrong. Similarly some air forces have been quite prepared to argue away empirical evidence for institutional advantage. That has skewed strategy and led to hundreds of thousands of ghastly deaths.

Consequently, the study of warfare should move away from excessive reliance on rationalistic argumentation. Discussing the discourse is not good enough. Writer 'A' may have said this; Writer 'B' may have written that; but so what? Relatively little is written about war*fare*. And when you next read something about warfare, ask yourself whether it actually comes down to anything more than appeals to the authority of some obscure academic(s), rhetoric, and assertion. The results of such rationalistic argumentation are not good enough. Some of the people who peddle them are charlatans. We see several of them in the early twenty-first century. Some are quite well-known. They may be well-meaning. But should we really put our soldiers', sailors' and airmen's lives at risk on the basis of consensus among academics? Or charlatans?

Surely not. We need to do better.

There are six broad historical issues of interest to warfare. They are:

1. what happened.
2. how it happened, in terms of sequence.
3. how it happened, in terms of mechanism.
4. why it happened, in terms of human agency (simplistically, 'why he (or she, or they) did it').
5. why it happened, in terms of tactical (or operational, or strategic) causation.
6. what we can learn from the above, to guide the conduct of future operations.

Historians are primarily interested in 1, 2 and 4. They provide context for the others. 6 is the main focus for future commanders in this area. 3 and 5 are the most useful. The history of warfare is mostly concerned with 3, 4 and 5. However 3 and 5 are grossly understudied. We observe that the academic-led pursuit of military history alone is inadequate to the task.

We do, however, have reason to be optimistic. We can now identify and discuss the strategic, operational and tactical levels of war in a way that our forebears in 1900 could not. We can now apply Clausewitzian thinking and consider ends, ways and means at each of those levels. We can now consider several theories to have been useful but now having shortcomings. They include those of Fuller, Liddell Hart, Boyd and Warden. We can now identify good examples of limited inductionism. Examples are the Seeckt reforms, Soviet deep battle theory and Slessor's writings. All this points towards a better future.

INDEX

www.ingramcontent.com/pod-product-compliance
Lightning Source LLC
LaVergne TN
LVHW051058210926
848486LV00002B/142